Palgrave Advances in Behavioral Economics

Series Editor
John Tomer
Co-Editor, Journal of Socio-Economics
Department of Economics & Finance
Manhattan College
Riverdale, NY, USA

This ground breaking series is designed to make available in book form unique behavioral economic contributions. It provides a publishing opportunity for behavioral economist authors who have a novel perspective and have developed a special ability to integrate economics with other disciplines. It will allow these authors to fully develop their ideas. In general, it is not a place for narrow technical contributions. Theoretical/conceptual, empirical, and policy contributions are all welcome.

More information about this series at
http://www.palgrave.com/gp/series/14720

Li Way Lee

Behavioral Economics and Bioethics

A Journey

palgrave
macmillan

Li Way Lee
Wayne State University
Detroit, MI, USA

Palgrave Advances in Behavioral Economics
ISBN 978-3-319-89778-3 ISBN 978-3-319-89779-0 (eBook)
https://doi.org/10.1007/978-3-319-89779-0

Library of Congress Control Number: 2018938323

Printed on acid-free paper

This Palgrave Pivot imprint is published by the registered company Springer International Publishing AG part of Springer Nature
The registered company address is: Gewerbestrasse 11, 6330 Cham, Switzerland

Preface

Grace Loo (MD, Shanghai Medical School) inspired me to take this journey. A long time ago, when I noticed that she was seeing some patients for free, I would ask her how much money she could make on that day. She would look at me in the eye and say: "Only so much as to buy food for the family." In those days, that meant 80 Taiwanese yens, or about 2 American dollars a day. She had a family of four; she was my mother.

Morris Altman was the editor of *Journal of Socio-Economics* when I began to send him manuscripts on behavioral bioethics. He would begin his comments with a four-word statement: "The paper makes sense." That was enough to spur me on in the direction of bioethics.

John Tomer, who encouraged me to write this book, has a wonderful, willful blindness to my inexperience as a book writer. He would call me every few months to find out if I had done anything. I dragged my feet until one day when I became convinced that he was really willfully blind. His kind and encouraging words still ring in my ears today.

Albert Lee keeps me grounded in the real world of bioethics. As I write this, I am reading what he is reading: *Addressing Patient-Centered Ethical Issues in Health Care: A Cased-Based Study Guide*, published by American Society for Bioethics and Humanities. I will ask him to be my health-care agent. He will do a fine job.

I am grateful for two anonymous reviews of the book prospectus. Both reviews were twice as long as the prospectus. I benefitted from both immeasurably. I am also grateful to John Breen, a student of mine, for volunteering to serve as a "test reader." A few days after I had sent him a draft, he ran into me and exclaimed: "It is easy to read! I wish people would write textbooks like that." At that moment, I knew that I was on the right track. John is an avid reader of Strunk and White's *Elements of Style*.

Last but not least, I thank two publishers. To Oxford University Press: for the permission to use the whole of my article:

Lee, Li Way, "International Justice in Elder Care: The Long Run," *Public Health Ethics*, 4(3), 2011(b), pp. 292–296.
To Elsevier: for permissions to use large parts of five of my articles:
Lee, Li Way, "The Predator-Prey Theory of Addiction," *The Journal of Behavioral Economics*, 17(4), Winter 1988, pp. 249–262.
Lee, Li Way, "Compassion and the Hippocratic Oath," *Journal of Socio-Economics*, 37(5), October 2008, pp. 1724–28.
Lee, Li Way, "Living Will: Ruminations of an Economist," *Journal of Socio-Economics*, 38(1), January 2009, pp. 25–30.
Lee, Li Way, "The Oregon Paradox," *Journal of Socio-Economics*, 39(2), April 2010, pp. 204–208.
Lee, Li Way, "Behavioral Bioethics: Notes of a Behavioral Economist," *Journal of Socio-Economics*, 40, August 2011(a), pp. 368–372.

Detroit, USA Li Way Lee

Contents

List of Figures

List of Tables

CHAPTER 1

Introduction

Abstract In this book I take a short journey through the universe of bioethics. I go two ways: inward and outward. By going inward, I see inner selves. They deal with many bioethical issues. By going outward, I find that we are linked to other entities in matters of bioethics, too. The universe of bioethics is limited only by our own perception.

Keywords Bioethics · Ethics · Justice · Dynamic justice
Static justice

1 The Universe of Bioethics

Bioethics is about living, dying, and death. By that definition, the universe of bioethics is very big indeed, as many things other than people live and die, too. I read somewhere that the universe ought to include biosphere. I like that vision very much. Still, I see the universe of bioethics as even bigger. I think that Future Earth, which is everything that lives and dies in the future, ought to be in that universe. Also, the universe extends not just outward from me, but also inward from me. A great analogy is quantum physics: the world of elementary particles is as big, if not bigger, than the rest of the physical universe.

So there are a lot of places to visit in the universe of bioethics. In this book I take a very short journey through that universe. I look

L. W. Lee, *Behavioral Economics and Bioethics*,
Palgrave Advances in Behavioral Economics,
https://doi.org/10.1007/978-3-319-89779-0_1

Fig. 1 My itinerary

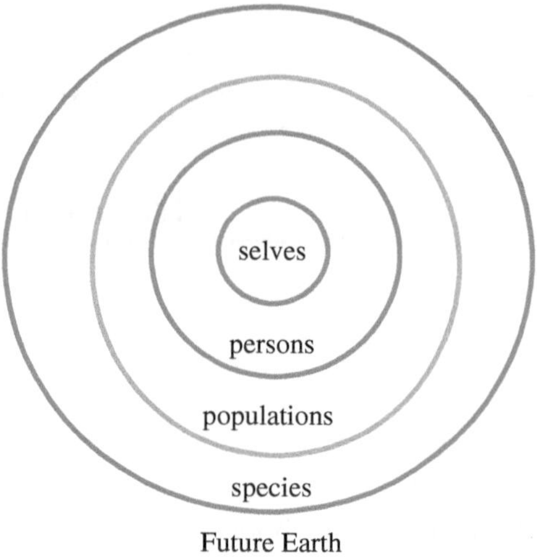

two ways: inward and outward. By looking inward, I see the mindset of a person. Our mind thrives with bioethics issues. (It is where a lot of behavioral economics comes into play.) By looking outward, I find that we are linked to other entities in matters of living, dying and death. The links are everywhere I look.

Looking both inward and outward, I arrange matters of life and death into four rings: selves, persons, populations (of people), and species. Figure 1 is a picture of the arrangement.

I wander into each of these rings. In each ring I make two or three stops. At each stop I look around and record whatever justice I see. Then, for the last stop of the journey, I step out of the rings and into Future Earth.

2 Justice and Ethics

At every stop in the journey, I look for the ethics that we do, not the ethics that we say. I will call "the ethics that we do" simply justice. Ethics that we say and justice are two different things. Here is a very short story:

> Two robbers, who do not know each other, are out looking for victims. Then one robber robs the other robber.

Robbery is unethical. But what happened to the two robbers in this story is justice served. Most of us would say: "They deserve each other." Here is another very short story that shows that an action may be ethical, but the underlying relation is unjust.

> The hunter corners the elephant, and then carefully aims the gun at the elephant's head. The hunter pulls the trigger, and the elephant falls to the ground and dies instantly.

You might say: "But that does not justify the killing!" I totally agree. The hunter is compassionate by aiming the gun at the elephants' head; nonetheless, the relationship between the hunter and the elephant is unjust to begin with.

Don't take me wrong. I love to read books on bioethics principles. Still, in the hustle and bustle of an ordinary day, few of us stop to ask if the things we do are ethical or not. We don't carry a list of "Five Moral Principles" in our pocket and check it every time we do something. That is understandable. We do most things out of habits; we don't question what we do every day. Even if we should question what we do, few of us would have the answers. What does an ethical action mean anyway?

3 Justices: Static and Dynamic

In my journey, I find two kinds of justice: static and dynamic.

Static justice prevails whenever parties are free to bargain with each other. Static justice is best captured by the Nash Solution (Luce and Raiffa 1957). The Nash Solution basically says: Let's meet in the middle. That means dividing equally the good thing to which we do not already have before we cooperate but we can have if we cooperate.[1]

[1] If there are 10 units of the good thing to share, and if you are already entitled to 3 and I to 1 before bargaining begins, then we would share equally what is left after accounting for these entitlements, or 6 (= 10 − 3 − 1). By dividing 6 in half, each of us would gain 3. Therefore, the 10 units are divided into 6 (= 3 + 3) for you and 4 (= 1 + 3) for me.

Dynamic justice follows from interactions over time, whether or not the parties bargain. Herbert Simon (1982) uses ecological models to explain this. I will follow him. In several of my visits I am struck by one phenomenon: an attempt to tip "the balance of justice" at any moment has a tendency to backfire. For example, I have seen that, when people raise more pigs to eat, both pigs and people will get sicker and die earlier, so that in the end there will be fewer pigs and fewer people. That is dynamic justice. It is subtle; it is evident only over time, often a very long time.

I report my findings in this book. I hope you enjoy them.

References

Luce, R. Duncan, and Howard Raiffa. *Games and Decisions.* New York: Wiley, 1957.

Simon, Herbert A. *Models of Bounded Rationality, Behavioral Economics and Business Organizations*, volume 2. Cambridge: MIT Press, 1982.

PART I

Selves in a Patient

CHAPTER 2

The Patient Who Changes His Mind

Abstract Bioethics should adopt the more nuanced view of rationality from behavioral economics. Most of us are conscious and capable of making decisions, but we are not consistently rational about all the issues all the time in all phases of life. And it is not to be taken for granted that we like to make decisions, whatever they are and whatever their consequences are. In this chapter, I make a case for bringing behavioral economics to bear on bioethics, so we have a bioethics that recognizes bounded rationality. I believe that such a "behavioral bioethics" will benefit both physicians and patients by bringing them together.

Keywords Behavioral bioethics · Patient · Time inconsistency
Bounded rationality

1 The Patient in Modern Bioethics

Modern bioethics is all about how to take care of the patient: What is good for the patient and what are the right ways of relating to the patient. In modern bioethics, the patient is supposed to be clear-headed, far-sighted, informed, and eager to made decisions. In other words, the patient is "mentally competent."

This chapter is adapted from Lee (2011).

L. W. Lee, *Behavioral Economics and Bioethics*,
Palgrave Advances in Behavioral Economics,
https://doi.org/10.1007/978-3-319-89779-0_2

However, mental competency proves to be a difficult concept. Bioethicists spend a lot of time debating what it means and making sense of it case by case (American Society for Bioethics and Humanities 2017). The fact is that most of us are, strictly speaking, not totally competent and some of us are more competent than others. That is how we are with respect to problems we are trying to solve in various phases of life (Veatch 2003; The President's Council on Bioethics 2005; Veatch et al. 2010). Yet, there is much pressure on health professionals to apply a dichotomy: competent or incompetent. Not surprisingly, this dichotomy works as well as a Procrustean bed.[1]

Bioethics should adopt the more nuanced view of rationality from behavioral economics. Most of us are conscious and capable of making decisions, but we are not consistently rational about all the issues all the time in all phases of life. And it is not to be taken for granted that we like to make decisions, whatever they are and whatever their consequences are. In this chapter, I make a case for "behavioral bioethics": a bioethics that recognizes bounded rationality. I believe that behavioral bioethics will benefit both physicians and patients by bringing them together again.[2]

2 The Patient Who Is Time-Inconsistent

Time inconsistency is a widely recognized trait among us. Time inconsistency means that my view of things, including preference, changes over time. For example, the older I grow, the more I wonder what I was thinking when I smoked cigarettes. Since smoking has been the cause of my health problems, it seems to me today to be a selfish behavior on the part of the young me. Other examples of time inconsistency are young people's disinterest in saving for old age (Akerlof and Shiller 2009, Chapter 10; Sunstein and Thaler 2003; Thaler and Benartzi 2004), their greater impatience (Read and Read 2004), and hyperbolic discounting (Laibson 1997).

Time inconsistency becomes more prominent over time. Our mind is better at solving problems in the short run than in the long run. We have beaten other species at managing a short life (say, 20 years), but we are

[1] Veatch (2003, p. 105) says that mentally incompetent patients put bioethics in a state of "moral chaos."

[2] A large literature loosely known as "behavioral ethics" (Trevino et al. 2006) explains how people come to behave more or less ethically.

still new at the game of managing a long one (say, 90 years). Our mind is not built to last 100 years (Nesse 2005, pp. 904–905). When it gets old, it is prone to work in odd ways. Half of the people older than 85 years show "cognitive deficit" (Lynn and Adamson 2003, p. 14). Time inconsistency is increasingly an issue among old people.[3]

One way of capturing time inconsistency is to think of a person as consisting of multiple selves. Smokers regularly resolve never to smoke again, only to smoke regularly again. A lot of us do a lot of other things that have bad long-term consequences, as if these consequences will befall others only (Frederick et al. 2004, pp. 190–191).[4] Schelling (1984, p. 112) wonders why people have a hard time deciding when and how to die: "there is no graver issue for the coming century than how to recognize and authenticate the preferences of people for whom dying has become the issue that dominates their lives." In his explanation, the two selves of Mr. Jones—the young Jones and the elderly Jones—assert themselves in turn, in different phases of life. The central issue is which self is authentic. If the young Jones signed a no-resuscitation order, but years later the elderly Jones wants to rescind it, then who is telling the truth? In Fig. 1, which head is authentic?

Posner (1995, pp. 84–94) wonders why there are conflicts between young people and old people. A person seems to acquire a different outlook in later life; the old self seems to become "a stranger to her younger self."

Bioethicists are aware of the two-self model and have made use of it. For example, in a report on old age, the President's Council on Bioethics (2005, p. 194) argues that a living will that instructs the physician to withhold all invasive treatments "discriminates against an imaginary

[3]In 1900, the average life expectancy of Americans was 47 years, while today it stands at nearly 78 years (Heron et al. 2009). This is an amazing change in 100 years, when Homo sapiens are said to have roamed the Earth for at least 100,000 years before this century. Another reason is that all phases of life, young and old, are getting longer. The President's Council on Bioethics puts it this way (2005, pp. 6–7, emphasis theirs): "*The defining characteristic of our time seems to be that we are both younger longer and older longer*...." The *later life* is most dramatically prolonged. Those who have lived to be 65 years old can expect to live to be 83.5 years old on average. And those who have lived to be 85 years old can expect to live to be 91.4 years old.

[4]Philosophers are interested in two-self models as well. See, for example, Parfit (1984), Daniels (1988), Buchanan and Brock (1989), and Dresser and Robertson (1989).

Fig. 1 Patient Jones

future self long before the true well-being of that future self is really imaginable." The young self has no knowledge of what it is like to be the old self (especially one who is mentally incompetent), and, therefore, should not be in a position to orchestrate the demise of the old self.

3 Bioethics and the Time-Inconsistent Patient

To argue my case for behavioral bioethics, I look at how principles of modern bioethics would apply when the patient is time-inconsistent.

3.1 The Principle of Nonmaleficence

The principle of nonmaleficence instructs a physician to avoid an action that, in the physician's best judgment, can cause harm to the patient. It seems to be a relatively easy principle to apply. In fact, however, it works like a straitjacket on the physician.

Consider the case *Natanson v. Kline* (Veatch 2003, p. 75). Natanson was 35 years old when she had surgery to remove breast cancer. She then had cobalt radiotherapy, a relatively new technology at the time. The radiation caused extensive damage to her skin, eventually the loss of use of a lung and an arm. She sued her physician, Dr. Kline, for not having informed her of the risks of the radiation and obtained her consent to the treatment. Dr. Kline claimed that doing so would likely have led Natanson to reject the therapy, and greater harm would have followed.

Obviously, Dr. Kline did hurt the 35-year old Natanson. But, at the same time, Dr. Kline believed that he saved a life: that of the future Natanson. Dr. Kline must have felt terribly constrained by the principle of nonmaleficence when he saw two Natansons: the young, who objects to radiation, and the old, who pleads for her life. Given that, the question that pressed Dr. Kline becomes a different one: what is the fair thing to do? The two-self model does not resolve the dilemma facing Dr. Kline, but it gives hope that the dilemma can be resolved if there is a way of judging if an action is fair.[5]

3.2 *The Principle of Autonomy*

The rise of this principle is widely regarded as the most significant development in bioethics in recent decades. Autonomy is, essentially, free choice. The principle of autonomy tells the physician to respect the choice of an informed patient. This seems to be totally reasonable, until we realize how little an ordinary patient can possibly know about medicine and until we find ourselves in disagreement with the patient's choice.

This dilemma can be illustrated by living wills.[6] A living will, from the point of view of rationality, seems to be a wise thing to have. When one is at the end of life, one may not appreciate being kept alive by a feeding tube. A living will directs that resources are not to be used for artificial feeding. With the assurance that the instruction will be followed, one may then leave the resources so saved to one's favorite charities or shift them to younger days when one can enjoy them more. Most bioethicists do not object to living wills. Some, however, do. In the above-mentioned report on old age, the President's Council on Bioethics asks (2005, p. 84): "(D)o we possess a present right to discriminate against the very life of a future self, or – even more problematic – to order others to do so on our behalf?" Here, the Council is using a two-self model, thereby admitting time inconsistency. Further, in its discussion of the "conceptual and moral limits" of living wills, the Council implies that the young self dominates the old self (see the President's Council, 2005, esp., p. 194). In this unequal relationship, there exists the potential of abuse of the old self by the young self, as the young self tries to minimize the transfer of income to the old self.[7] Suffice it to say that not all bioethicists embrace complete patient autonomy.

[5]Mrs. Natanson lived well into the sixties and died of cancer of an unknown origin (*Breast Cancer Action*, Newsletter 83, Fall 2004).

[6]For the purposes of this chapter, an advance directive and a living will are interchangeable terms. Technically, an advance directive consists of two parts: a living will and a healthcare proxy.

[7]Suspicion that one self can exploit the other underlies laws against suicide. A young self who commits suicide effectively prevents the old self from coming into existence.

Indeed, much tension has been building within the principle of autonomy. The tension is responsible for the moral quagmire in which many elderly find themselves. These elderly were "previously competent" but can no longer understand choices. Therefore they are no longer autonomous. But what if their living wills, drafted while they were competent, spell out clearly what is to be done and what is not to be done in case they are judged to be incompetent? In practice, their physicians and even their surrogates have been less than willing to honor their living wills. There is much resistance to "extending autonomy" to elderly people with dementia. This is best illustrated, again, by a "hard case" in the 2005 report of the President's Council on Bioethics. An Alzheimer's patient is discovered to have a malignant but operable tumor. The patient has a living will that states that "he wants no invasive treatments of any kind once his dementia has progressed to the point where he is no longer self-sufficient and can no longer recognize his family members" (President's Council 2005, p. 193). His daughter, who years ago promised to honor the living will, now wants to ignore it, since he seems "generally cheerful."[8]

So, under the weight of the evidence that patients are not consistently rational throughout life, the principle of autonomy—a cornerstone of modern bioethics—is showing cracks. It increasingly appears that the ascendancy of the principle of patient autonomy is a Pyrrhic victory. Both the patient and the physician will benefit from a bioethics that recognizes time inconsistency.

3.3 The Principles of Fidelity and Veracity

Fidelity and veracity are two other principles of ethical actions on the part of the physician. The principle of fidelity means that the physician is bonded to the patient by trust and must not break the bond no matter how tempting it is to do so. The principle of veracity means that the physician must tell the patient the truth, no matter how unpleasant it is to the patient or the physician. Thus, when a physician has agreed to

[8] Posner (1995, pp. 259–260) and Veatch (2003, p. 106) suggest that "the principle of autonomy extended" can be applied to a person who was previously competent but has fallen into a permanent vegetative state. A vegetative state is presumably inert, while a demented state—merely mysterious—is not inert.

accept a patient, the physician must remain loyal to the patient and must be truthful. These principles seem quite straightforward.

Yet, in reality, physicians often find these principles difficult to apply because the patient is not quite as simple as assumed. For example, must the physician remain loyal to the patient if the patient does not pay bills? Or, must the physician tell the patient the truth of his illness if the patient adamantly refuses to know the truth? Problems of moral hazard, self-denial, and strategic behavior are standard fare in behavioral economics, but they are treated as anomalies in bioethics.[9]

I consider principles of fidelity and veracity together because I want to make another point: that it is fairly easy for these principles to clash. In a case described also in Veatch (2003, pp. 81–82), Dr. Wordsworth is giving a routine physical exam to a young patient, Mr. Sullivan, who is obese and smokes a lot. Sullivan has made it clear that he has no intention of losing weight or smoking less. What is Wordsworth supposed to do? The principle of fidelity means that Wordsworth must do something; otherwise, Sullivan is likely to get ill, as if abandoned by Wordsworth. The principle of veracity means that Wordsworth must tell Sullivan the truth of the chest x-ray: it looks fine. In the end, Dr. Wordsworth decides to scare Sullivan into quitting smoking, by telling him that some spots on the x-ray suggest precancerous development in his lungs.

Clearly, Dr. Wordsworth has lied, in violation of the principle of veracity. However, if Wordsworth had not lied, then Sullivan would not have quit smoking, and Wordsworth would have violated the principle of fidelity. When he accepted Sullivan as a patient, Dr. Wordsworth made a promise to take care of his health, not only on that day but also in the future. It is as if there are two Sullivans. If Wordsworth did not try to stop the present Sullivan from smoking, then Wordsworth would be disloyal to the future Sullivan, who is likely to suffer from poor health in the body left behind by the young Sullivan.

Again, my point here is that bioethics ought to recognize that a patient can change his mind, as if there are two selves in Sullivan. Otherwise Dr. Wordsworth must choose to be a liar or a traitor. If bioethics could allow for a behaviorally more complex patient, then it would help the physician identify the appropriate ethical principle in dealing with the patient. To this possibility we now turn.

[9] See Veatch et al. (2010) for case studies.

4 Social Justice in the Time-Inconsistent Patient

When the physician makes a decision that affects several patients, the physician often must tussle with the question of social justice. For example, if there is only one kidney available for transplant today and there are two patients who need it, any decision will benefit one and not the other. Is it fair to give the kidney to the younger of two patients?

I shall leave this chapter with an observation: social justice also arises when a physician treats a single patient who is time-inconsistent. We saw it in the case of Dr. Kline not telling Irma Natanson that radiation has risks. We also saw it in the case of Dr. Wordsworth's twisted interpretation of a chest x-ray.

In a later chapter, I shall return to social justice by considering the extension of the life of a single patient. The story there is pretty simple. Suppose that a biomedical technology has been invented that extends the life of the *old* self and it is available to anyone at no cost. What is not to be happy about that? There are several concerns: first, with a longer life, the old self immediately faces a lowering of the standard of living; second, the old self will want to call on the young self to transfer income; third, the young self will feel unhappy about that, even while feeling obligated to maintain their relationship in some sort of balance. In any case, the young self is adversely affected by the old self's longer life, and will feel like being dealt a bad hand.

Therefore, before the kidney transplant, the physician will want to get to know the two selves in the young patient. As the patient may need time to let the internal negotiation reach a compromise, and as the patient may speak in two voices, the physician needs to take time to listen carefully.

References

Akerlof, George A., and Robert J. Shiller. *Animal Spirits.* Princeton: Princeton University Press, 2009.

American Society for Bioethics and Humanities. *Addressing Patient-Centered Ethical Issues in Health Care: A Case-Based Study Guide.* Chicago, 2017. https://www.asbh.org.

Buchanan, Allen E., and Dan W. Brock. *Deciding for Others: The Ethics of Surrogate Decision Making.* Cambridge: Cambridge University Press, 1989.

Daniels, Norman. *Am I My Parents' Keeper? An Essay on Justice between the Young and the Old.* New York: Oxford University Press, 1988.

Dresser, Rebecca S., and John A. Robertson. "Quality of Life and Non-Treatment Decisions for Incompetent Patients: A Critique of the Orthodox Approach." *Law, Medicine, and Health Care*, 17 (3), 1989, pp. 234–244.

Frederick, Shane, George Loewenstein, and Ted O'Donaghue. "Time Discounting and Time Preference: A Critical Review." In Colin F. Camerer, George Loewenstein, and Matthew Rabin, eds. *Advances in Behavioral Economics*. New York: Russell Sage Foundation; Princeton and Oxford: Princeton University Press, 2004.

Heron, Melonie, et al. "Deaths: Final Data for 2006." *National Vital Statistics Reports*, 57 (14), April 2009.

Laibson, David. "Golden Eggs and Hyperbolic Discounting." *Quarterly Journal of Economics*, 62, May 1997, pp. 443–477.

Lee, Li Way. "Behavioral Bioethics: Notes of a Behavioral Economist." *Journal of Socio-Economics*, 40, August 2011, pp. 368–372.

Lynn, Joanne, and David M. Adamson. "Living Well at the End of Life: Adapting Health Care to Serious Chronic Illness in Old Age." RAND Health White Papers, Santa Monica, California, 2003.

Nesse, Randolph M. "Evolutionary Psychology and Mental Health." Chapter 23 in David M. Buss, ed. *The Handbook of Evolutionary Psychology*. Hoboken, NJ: Wiley, 2005.

Parfit, Derek. *Reasons and Persons*. Oxford: Clarendon Press, 1984.

Posner, Richard A. *Aging and Old Age*. Chicago and London: The University of Chicago Press, 1995.

President's Council on Bioethics. *Taking Care: Ethical Caregiving in Our Aging Society*. Washington, DC, September 2005. Available at http://www.bioethics.gov/reports/taking_care/index.html.

Read, Daniel, and N. L. Read. "Time Discounting over the Lifespan." *Organizational Behavior and Human Decision Processes*, 94, 2004, pp. 22–32.

Schelling, Thomas C. *Choice and Consequence*. Cambridge: Harvard University Press, 1984.

Sunstein, Cass, and Richard Thaler. "Libertarian Paternalism." *American Economic Review*, 93, 2003, pp. 175–179.

Thaler, Richard, and Shlomo Benartzi. "Save More Tomorrow™: Using Behavioral Economics to Increase Employee Saving." *Journal of Political Economy*, 112 (S1), 2004, pp. 164–187.

Trevino, Linda K., Gary R. Weaver, and Scott J. Reynolds. "Behavioral Ethics in Organizations: A Review." *Journal of Management*, 32 (6), December 2006, pp. 951–90.

Veatch, Robert M. *The Basics of Bioethics*, 2nd ed. Upper Saddle River, NJ: Pearson Education, 2003.

Veatch, Robert M., Amy M. Haddad, and Dan C. English. *Case Studies in Biomedical Ethics*. New York: Oxford University Press, 2010.

CHAPTER 3

The Two Selves in My Friend Addict

Abstract My friend Addict has two selves: good and bad. The bad self preys on the good self. Their interactions give rise to Addict's periodic feeling of conflict. I find that, to diminish the bad self, we cannot simply try to harass the bad self or favor the good self; we must begin by diminishing the good self. This remedy works like the scorched-earth tactic in a battle. Only by starving the bad self will we succeed in preserving the good self.

Keywords Addiction · Two selves · Predator self · Prey self

1 Introduction

I have heard two stories about addiction. In one, addiction is a decision (Stigler and Becker 1977; Becker and Murphy 1988). In the other, addiction is a continuing conflict between two selves.

In this chapter, I visit a close friend of mine, by the name of Addict, who complains about feeling conflicted all the time. Addict also tells me that his feeling of conflict changes periodically. That reminds me of Schelling's (1984, p. 59) accounts of such behavior. There is the smoker who "grinds his cigarettes down the disposal swearing that this time he

This chapter is based on Lee (1988).

L. W. Lee, *Behavioral Economics and Bioethics*,
Palgrave Advances in Behavioral Economics,
https://doi.org/10.1007/978-3-319-89779-0_3

means never again to risk orphaning his children with lung cancer and is on the street three hours later looking for a store that's still open to buy cigarettes." There is the glutton "who eats a high-calorie lunch knowing that he will regret it, does regret it, cannot understand how he lost control, resolves to compensate with a low-calorie dinner, eat a high-calorie dinner knowing he will regret it, and does regret it." Schelling (1984, p. 70) then observes that certain afflictions "occur cyclically, on a schedule that is physiological or that reflects the daily or weekly pattern of living, or on some cycle autonomous to the habit itself, a cycle of onset and exhaustion and recovery...."[1]

In this chapter, I try to make sense of the periodic behavior in my friend Addict. It seems to me that there are two selves in Addict. Call them the "bad" self and the "good" self. Addict smokes. When he gets up in the morning, Addict is the good self, cursing the urge for a smoke, but the next minute Addict is the bad self, fumbling for the lighter and counting how many Camel cigarettes are left in the pack.

In spite of the periodic change in behavior, Addict seems to have become reconciled to it. For the longest time, however, other persons want to intervene. In Addict, I find that these attempts have surprising consequences. Attempts to promote the good self will do the good self no good, though also no harm; they will only help grow the bad self. Attempts to diminish the bad self will not affect the bad self at all, though they will make the good self grow stronger. This is some sort of dynamic justice at work.

2 The Predator–Prey Model of Addiction

The host of my visit, Addict, has two selves: "the bad self" B and "the good self" G. The prominence of B is measurable, and so is that of G. The addictive substance being tobacco, B may be measured by nicotine

[1] Others see addiction in similar light. Thaler and Shefrin (1981, p. 105) see it as a platform for a far-sighted "planner" and a short-sighted "doer." Sen (1976) sees in it a "meta-ranking" of preferences of several selves. Etzioni (1986, p. 159) sees an addict as "at least two irreducible sources of value or 'utility,' pleasure and morality." Weil and Rosen (1983) see addiction as the body of two relations, one with a person and the other with a drug like heroin.

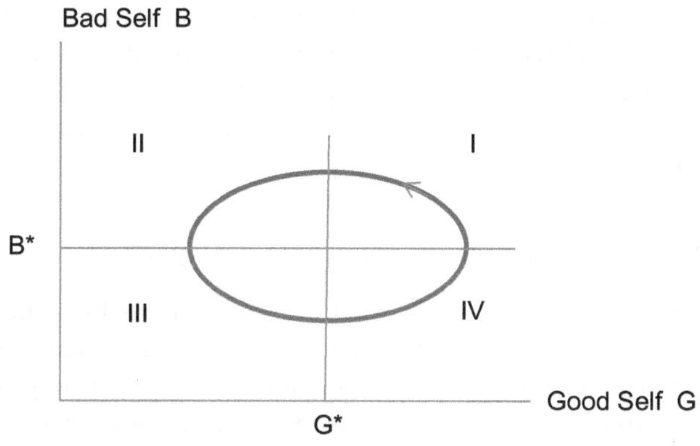

Fig. 1 My friend Addict

intake in 24 hours, and G by life expectancy. (If the addiction is gambling, then B may be measured by time spent on gambling, and G by net worth.)

The two selves relate to each other as predator and prey. Their "ecology" can be described by four processes:

1. The bad self subsists on the good self, growing at a rate proportional to the intensity of predation on the good self.
2. The bad self fades at a constant percentage rate. (So, if Addict is totally deprived of tobacco for a long time, then the desire for smoking will wear off.)
3. The good self grows steadily, also at a constant percentage rate.
4. The good self is drained by the bad self at a rate proportional to the intensity of predation by the bad self.

This predator–prey theory, also known as parasite–host theory, has found applications in economics (Boulding 1950; Hirshleifer 1977; Winston 1980). There are several versions. The version that I think applies to Addict best is pictured in Fig. 1.

Addict is constantly in motion, traveling counterclockwise on the loop. In phase I, the bad self is waxing while the good self is waning. In phase II, both selves decline. In phase III, the good self has turned the corner, while the bad self continues to decline. Then in phase IV the two selves both grow. Thus the two selves do not literally alternate in command. Their relative dominance changes continuously. The surrender of control is gradual, never complete. This oscillatory behavior is common among addicts.

3 Policies Against Addiction

For the longest time all over the world, addiction has been targeted as a social problem. Public policies toward it have mushroomed. The policies belong to two categories: "treatment" and "war" (Zinberg 1984; Lee 1993; Kristof 2017). In my story, these two categories correspond neatly to "promote the good self" and "demote the bad self."

How do these policies work in the predator–prey model? Not well at all, I am afraid. The reasons are counterintuitive and I am not sure how to explain them without hiding behind mathematics. Here I merely direct your gaze to Fig. 1. (You can find the math in the Appendix.) On the axis for the good self G, we see G*, which is the good self's average prominence over a cycle. The anomaly is that G* consists of characteristics of the *bad*—not the good—self. Furthermore, on the axis for the bad self B, we see B*, which is the bad self's average prominence over a cycle. The anomaly is that B* consists of characteristics of the *good*—not the bad—self. These anomalies have serious consequences.

3.1 Promote-the-Good-Self Policies

Our attempts to help the good self will do the good self no good, even as they will make the bad self grow stronger. Consider a mandatory fitness program for addicts. It increases the good self's rate of regeneration. Unfortunately, it will not affect the good self after all, though it will energize the bad self. Or consider low-tar cigarettes and methadone, which lower the bad self's drain on the good self. Again, in the long run, these substitutes have no effect on the good self; though they reinforce the bad self's desire for cigarettes and heroin. An intuitive explanation is that the bad self is parasitic, so he grows as the good-self host grows. Measures taken to improve the health of the good self will only support the habit of the bad self. The initial rise in health will be completely offset by the subsequent decline in health.

3.2 Demote-the-Bad-Self Policies

Similarly, our attempts to diminish the bad self will not affect the bad self. Consider the program "War on Drugs." It is a program aiming at the bad self, by making addictive substance more expensive to buy, more difficult to find, and more hazardous to procure. In the end, the war will fail to stunt the bad self. Incidentally, it will help the good self become stronger. That, however, is not the goal of the program.

4 Justice in Addiction

From my visit with Addict, I take home a conclusion that will irk many experts of addiction: The only way to diminish the bad self is to diminish the good self. That is, we must make my good friend Addict older, sicker, and weaker—anything that robs the good self in Addict of the ability to replenish and rejuvenate. We will succeed in culling the predator only by culling the prey.

Interestingly, this "scorched-earth" strategy will not scorch the earth after all: ultimately, the good self will recover fully after the bad self has been weakened.

I see justice in this paradox. The justice is that we will fail in our attempt to favor one self over the other self, whatever our prejudice toward them is. If we want to promote one, we must promote the other first; if we want to demote one, we must demote the other first.

Appendix

To understand addiction, I employ the simplest Lotka–Volterra model of predator-and-prey interactions (Kemeny and Snell 1972; Pielou 1969; Wilson and Bossert 1971; Smith 1974). The model consists of two sets of interactions between the bad self and the good self. The bad self grows with the intensity of predation on the good self and declines at a natural rate. The overall change over time, therefore, can be described by a linear differential equation:

$$dB/dt = b_1BG - b_2B \tag{1}$$

where t is time, b_1 is the coefficient of growth due to interaction, and b_2 is the rate of fading. They all have positive values. The good self is similarly governed by a differential equation:

$$dG/dt = g_1 G - g_2 BG \tag{2}$$

where g_1 is the rate of natural regeneration and g_2 is the coefficient of decline due to predation by the bad self. Both also have positive values. When the rate of regeneration of the good self exceeds the rate at which it is drained by the bad self, the good self will grow in strength. If it is the other way around, the good self will wither.

The main properties of the model are depicted in Fig. 1. The addict travels on a closed loop. Once around the loop, the good self has average presence of

$$G^* = b_2/b_1 \tag{3}$$

and the bad self has average presence of

$$B^* = g_1/g_2 \tag{4}$$

Oscillation is not a defining property of addiction, however. An addict may have a fixed, stable proportion of good self and bad self. This can result from a slightly different model of predator and prey. Suppose that an addict's good self rejuvenates at the "logistic" rate that is slightly lower than the exponential rate. If the good self is measured by health status, then the addict rejuvenates at less than the replacement rate. Then the good self and the bad self will converge toward a stable equilibrium. That equilibrium has the same "scorched-earth" property as that of the oscillating equilibrium. Details can be found in Lee (1988).

References

Becker, Gary, and Kevin Murphy. "A Theory of Rational Addiction." *Journal of Political Economy*, 96, August 1988, pp. 675–700.

Boulding, Kenneth E. *A Reconstruction of Economics.* New York: Wiley, 1950.

Elster, Jon, ed. *The Multiple Self.* New York: Cambridge University Press, 1986.

Etzioni, Amitai. "The Case for a Multiple-Utility Conception." *Economics and Philosophy*, 2, 1986, pp. 159–183.

Hirshleifer, Jack. "Economics from a Biological Viewpoint." *Journal of Law and Economics*, 20, April 1977, pp. 1–52.

Kemeny, John G., and J. Laurie Snell. *Mathematical Models in the Social Sciences.* Cambridge: MIT Press, 1972.

Kristof, Nicholas. "How to Win a War on Drugs." *New York Times*, 22 September 2017. Accessed at https://www.nytimes.com/2017/09/22/opinion/sunday/portugal-drug-decriminalization.html?action=click&pgtype=

Homepage&clickSource=story-heading&module=opinion-c-col-left-region&region=opinion-c-col-left-region&WT.nav=opinion-c-col-left-region.
Lee, Li Way. "The Predator-Prey Theory of Addiction." *The Journal of Behavioral Economics*, 17 (4), 1988, pp. 248–262.
Lee, Li Way. "Would Harassing Drug Users Work?" *Journal of Political Economy*, 101, October 1993, pp. 939–959.
Pielou, E. C. *An Introduction to Mathematical Ecology*. New York: Wiley, 1969.
Schelling, Thomas C. *Choice and Consequence*. Cambridge, MA: Harvard University Press, 1984.
Sen, Amartya K. "Rational Fools: A Critique of the Behavioural Foundations of Economic Theory." *Philosophy and Public Affairs*, 6, 1976–1977, pp. 317–344.
Smith, J. Maynard. *Models in Ecology*. London: Cambridge University Press, 1974.
Stigler, George J., and Gary S. Becker. "De gustibus non est disputandum." *The American Economic Review*, 67, March 1977, pp. 76–90.
Thaler, Richard H., and H. M. Shefrin. "An Economic Theory of Self-Control." *Journal of Political Economy*, 89, April 1981, pp. 392–406.
Weil, Andrew, and Winifred Rosen. *Chocolate to Morphine: Understanding Mind-Active Drugs*. Boston: Houghton-Mifflin, 1983.
Wilson, Edward O., and William H. Bossert. *A Primer of Population Biology*. Stamford, CT: Sinauer Associates, 1971.
Winston, Gordon C. "Addiction and Backsliding." *Journal of Economic Behavior and Organization*, 1, December 1980, pp. 295–324.
Zinberg, Norman E. *Drug, Set, and Setting*. New Haven: Yale University Press, 1984.

CHAPTER 4

The Oregon Paradox

Abstract When terminally ill people are given the option of legally *hastening* death, they often feel a sense of greater well-being and a desire to live *longer*. In my explanation of this paradox, a terminally ill person has two selves. The right-to-die empowers the future self to gain control of suffering at the end of life. That makes the present self, who has empathy with the future self, feel a surge in well-being and the desire to live a longer life.

Keywords Right to die · Death with Dignity Act · Present self
Future self · Well-being

1 DWDA and the Paradox

In 1997, the state of Oregon passed the Death with Dignity Act (DWDA). A resident there who has been certified to have no more than 6 months of life left may choose to die by taking barbiturates such as Secobarbital and Pentobarbital, morphine, or combinations thereof. The prescription must be written by a physician and filled by a pharmacist.

DWDA patients, upon obtaining the lethal medicine, often feel a surge in well-being and peace of mind, but also a desire to live *longer*.

This chapter is largely based on Lee (2010).

L. W. Lee, *Behavioral Economics and Bioethics*,
Palgrave Advances in Behavioral Economics,
https://doi.org/10.1007/978-3-319-89779-0_4

This phenomenon is paradoxical since these feelings stem from an ability to end life *sooner*:

> **The Oregon Paradox**: When terminally ill people are given the option of legally *hastening* death, they often feel a sense of greater well-being and a desire to live *longer*.

Note that there are two parts to the paradox: the enhanced feeling of well-being and the desire to live longer.

Below are nine striking anecdotes. They are narratives by daughters, physicians, and others who were close to those who died.

1. Lester Angell, 81 years old, fell and might have broken a bone. The day before he was scheduled to be taken to the hospital, he shot himself. His daughter, Marcia Angell, former Editor-in-Chief of *New England Journal of Medicine*, says: "If he knew he had the option to get help in ending his life at any time in the future, he probably also would have chosen to live longer" (Angell 2004, p. 21).
2. Anna, with ovarian cancer, obtained the medicine and said: "I felt I had more energy to fight the cancer and just to live in the present time. It just took a big weight off my shoulders somehow, knowing at least that that was one thing that maybe I didn't have to worry about" (Pearlman and Starks 2004, pp. 92–93). She did not take the medication until 3 years later.
3. Diane, diagnosed with leukemia, refused treatments. She obtained barbiturates with the help of her physician, who observed: "It was extraordinarily important to Diane to maintain control of herself and her own dignity during the time remaining to her. When this was no longer possible, she clearly wanted to die" (Quill 1991, p. 693). She lived for several more months before taking the medicine.
4. Jim Romney, a plaintiff in *Oregon v Ashcroft* (2001) and an avid fisherman, suffered from Lou Gehrig's disease. He obtained the medicine under DWDA in 2002 and died in 2003. He did not use the medicine, but he said of DWDA: "…just knowing that this law is an available option is very liberating.… I feel so liberated today that I may go out and catch a Chinook salmon on the Columbia tomorrow" (Coombs Lee 2003, p. 3).

5. Penny Schleuter, an economist suffering from ovarian cancer, said the day before she took the medicine: "I want the security of knowing the option is there. After Attorney General Janet Reno said the Drug Enforcement Agency should not get involved, that it was a states' rights issue, I felt great comfort. Now I am filled with contentment and peace I simply did not have when the law was tied up in the courts" (Coombs Lee, p. 31).
6. James, a psychiatrist with pancreatic cancer, did not try to get the medicine. But he said that "it was liberating to know his death could be in his control" (Coombs Lee, p. 51).
7. Marcia, with malignant brain tumors, said that "it gave her serenity to know she could (get the medicine)" even though she did not try to get it (Coombs Lee, p. 54).
8. Charles, with lung cancer, obtained the medicine. Then he said: "Now that I have my security stash, I've decided to live until I die." He never used the medicine and died of cancer 3 months later (Coombs Lee, p. 54).
9. Richard Holmes, who had cancer, was the first plaintiff in *Oregon v Ashcroft* (2001). He died in 2002 without using the medicine that he fought for. His daughter, Sandy, later wrote: "…I'm certain he was comforted just knowing he had the drugs. I can remember the day he got the prescription filled. We talked on the phone that day and I could just hear the change in his voice. He felt much more in control. He knew that he had power over his life again, and after all he'd been through, it was exactly what he wanted" (Coombs Lee, p. 92).

For other evidence of the Oregon Paradox, we may note that hundreds of Oregonians took the considerable trouble of getting the medicine. Each of them made three requests (two verbal and one written), had two physicians and possibly a psychiatrist certifying terminal illness and mental competency, and then waited at least 15 days before getting the medicine.[1] They would not have bothered if they did not feel better off having the medicine in their hands.[2]

[1]To qualify, a resident must make three requests (two verbal and one written), have two physicians and possibly a psychiatrist certify terminal illness and mental competency, and then wait at least 15 days before getting the medicine. See Oregon's website for detailed instructions.

[2]Many more residents initiated the application, but did not follow through with it (Tolle et al. 2004).

In this chapter, I will tell a story about how the surge in well-being comes about. DWDA works on our emotions much like auto insurance: It sets a floor to loss. A terminally ill patient is a lottery with two outcomes: "good" and "bad." What DWDA provides is the option of cutting loss if the outcome turns out to be bad. This story is straightforward and often has been told by those who are terminally ill (Coombs Lee 2003, p. 23; Preston 2007, p. 132).

I will also tell a similar story about why we want to live a *longer* life when DWDA gives us the option of living a *shorter* life. Again, I make use of the lottery metaphor, with the good outcome and the bad outcome. If DWDA eliminates the worst consequences of the bad outcome, then it is as if we have improved the odds of the good outcome. And that gives us greater hope that we will live a longer, happier life.

2 The Good Path and the Bad Path

People die in different ways: some accidentally, some following a period of illness. Among those who get ill first, some die quickly, and some slowly.

Figure 1 describes two alternative trajectories of a terminally ill person's total utility over time: "the good path" and "the bad path."[3] Total utility is a gauge showing what a person feels at any point in time about living out the rest of life. Total utility can be negative, becoming "disutility." Utility, therefore, can show pain and indignity (Hamermesh and Soss 1974; Posner 1995, p. 255; Yang and Lester 2006, pp. 547–548).

Let's say that a person becomes terminally ill and learns of the diagnosis. At that point, the person sees two "alternative selves," one travelling down the "good path" and the other down "the bad path." On the good path, the person is able to enjoy a fairly constant level of utility until death comes suddenly. On the bad path, the quality of life declines steadily and relentlessly, each day worse than the day before. At some point on the bad path, there is no qualify of life left, utility becomes zero, and life is not worth living beyond that.[4]

[3]These paths are based on well-known "trajectories of chronic illness" (Lynn and Adamson 2003; Gawande 2014, Chapter 2; Zitter 2017, Appendix One).

[4]My model is similar to the one that Posner (1995, pp. 245–250) uses in his analysis of suicide. Posner also assumes that a person faces two possible future states: "the doomed state" and "the healthy state." In the doomed state, a person's utility is negative.

Fig. 1 Paths of dying

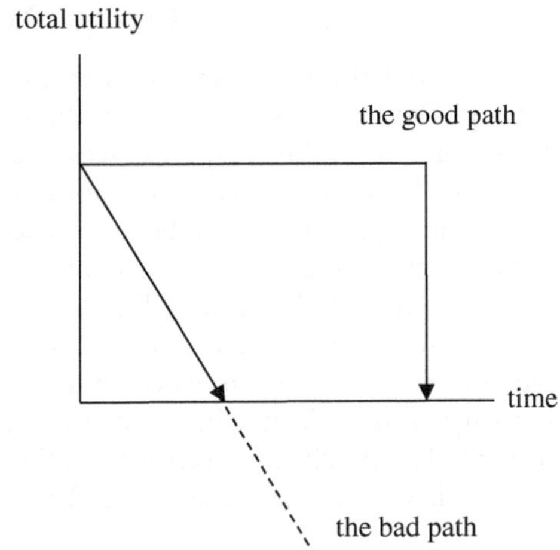

3 The Paradox Explained

Imagine that a patient feels the average of two total utilities: one on the good path and one on the bad path. Economists will recognize this as a form of "expected utility": If expected utility is positive, one wants to live; if negative, one does not want to live. To form expected utility, the patient will need to have estimates of the probabilities of the two paths. The patient will get these estimates from the physicians who provide the diagnosis.

With DWDA, the patient still faces the two paths, except that the bad path now is truncated. In Fig. 1, the dashed-line portion of the bad path is gone because the patient who uses DWDA will not be on that portion of the bad bath. That is a good thing because the dashed portion of the bad path is associated with negative utility.

The truncation of the worst part of the bad path explains the surge of well-being of the patient who chooses to use DWDA. The self on the good path is no worse off while the self on the bad path is better off because there is no longer a period of negative utility. So the patient who chooses DWDA is better off.

The truncation also explains the second part of the paradox: The patient wants to live longer with DWDA than without. When one can avoid the possibility of suffering at the end, one's average total utility over time cannot but go up at the end. So the patient wants to postpone the end—that is, to live a longer life.

4 The Other Oregon Paradox

When I told the story several years ago in a journal, a reader asked me why, if DWDA is such a good thing, more Oregonians do not take advantage of it. My response was that perhaps the explanation is the same as why so many people do not enroll in 401(k) retirement plans.

Or the explanation may be as simple as the endowment effect (Thaler 1980; Kahneman et al. 1991). People today do not see themselves as in full control of the whole range of life—death and dying, for example, are heavily regulated by laws and customs—and they do not feel in full ownership of old age. With DWDA, however, a life-and-death decision is literally placed in their hands. This may have a profound psychological effect on the terminally ill. The psychiatrist Terman, finding the Oregon Paradox to be "life's greatest irony," explains this effect (2007, pp. 158–159; his emphases):

> Knowing they can choose the time to permanently end their suffering provides them an opportunity to change their attitude toward their symptoms. One day at a time, they can decide if their lives still retain sufficient meaning to endure their suffering. If they have ultimate control over ending their lives but choose not to do so, then they have **voluntarily decided to stay alive despite their symptoms**. Thus patients can make an important psychological shift in how they see themselves: from "**helpless victims**" **to "willing survivors.**" …This positive change in self-perception, plus the decrease in anxiety about how intense and long suffering could be, permits patients to direct their energy toward a final search for meaning during the last months or weeks of their lives.

In other words, DWDA causes a shift in our anchoring point, which determines the way we view gains and losses of life. Without the medicine, everyday we live is a gift from God, which we cannot take for granted. If we die, it is because God has decided that our time is up and we cannot complain. With the medicine in our hands, however, much is changed: we view every day we live as a gain for which we are responsible, and every day we do not live as a loss for which we are also responsible. We own our life now. Given our aversion to loss, we have another reason for wanting to live longer.[5]

[5] We may go even a step further. Knowing that the medicine is available at the local pharmacy for the asking, we may not bother to get it—since we know we probably would not use it.

Death requires people to make momentous decisions under unfamiliar circumstances and extreme uncertainty. This is the kind of environment in which behavioral theories have a proven comparative advantage.

References

Angell, Marcia. "The Quality of Mercy." In Timothy Quill and Margaret P. Battin, eds. *Physician-Assisted Dying: The Case for Palliative Care and Patient Choice.* Baltimore: The Johns Hopkins University Press, 2004, pp. 15–23.

Coombs Lee, Barbara, ed. *Compassion in Dying: Stories of Dignity and Choice.* Troutdale: New Sage Press, 2003.

Gawande, Atul. *Being Mortal: Medicine and What Matters in the End.* Farmington Hills, MI: Gale Cengage Learning, 2014.

Hamermesh, Daniel S., and Neal M. Soss. "An Economic Theory of Suicide." *Journal of Political Economy*, 82 (1), 1974, pp. 83–98.

Kahneman, Daniel, Jack L. Knetsch, and Richard H. Thaler. "The Endowment Effect, Loss Aversion, and Status Quo Bias." *Journal of Economic Perspectives*, 5 (1), 1991, pp. 193–206.

Lee, Li Way, "The Oregon Paradox." *Journal of Socio-Economics*, 39 (2), April 2010, pp. 204–208.

Lynn, Joanne, and David M. Adamson. "Living Well at the End of Life: Adapting Health Care to Serious Chronic Illness in Old Age." RAND Health White Papers, Santa Monica, California, 2003.

Pearlman, Robert, and Helene Starks. "Why Do People Seek Physician-Assisted Death?" In Timothy E. Quill and Margaret P. Battin, eds. *Physician-Assisted Dying: The Case for Palliative Care and Patient Choice.* Baltimore: The Johns Hopkins University Press, 2004, pp. 91–101.

Posner, Richard A. *Aging and Old Age.* Chicago: The University of Chicago Press, 1995.

Preston, Tom. *Patient-Directed Dying: A Call for Legalized Aid in Dying for the Terminally Ill.* New York: iUniverse Star, 2007.

Quill, Timothy E. "Death and Dignity: A Case of Individualized Decision Making." *New England Journal of Medicine*, 324, 1991, pp. 691–694.

Quill, Timothy E., and Margaret P. Battin, eds. *Physician-Assisted Dying: The Case for Palliative Care and Patient Choice.* Baltimore: The Johns Hopkins University Press, 2004.

Terman, Stanley A. *The Best Way to Say Goodbye: A Legal Peaceful Choice at the End of Life.* Carlsbad: Life Transitions Publications, 2007.

Thaler, Richard H. "Toward a Positive Theory of Consumer Choice." *Journal of Economic Behavior and Organization*, 1, 1980, pp. 39–60.

Tolle, Susan W., Virginia P. Tilden, Linda L. Drach, Erik K. Fromme, Nancy A. Perrin, and Katrina Hedberg. "Characteristics and Proportion of dying Oregonians Who Personally Consider Physician-Assisted Suicide." *The Journal of Clinical Ethics*, 15 (2), 2004, pp. 111–122.

Yang, Bijou, and David Lester. "A Prolegomenon to Behavioral Economic Studies of Suicide." In Morris Altman, ed. *Handbook of Contemporary Behavioral Economics*. Armonk, NY: Sharpe, 2006, pp. 543–559.

Zitter, Jessica. *Extreme Measures: Finding a Better Path to the End of Life*. New York: Avery, an Imprint of Penguin Random House, 2017.

PART II

Physician and Patient

CHAPTER 5

The Two-Headed Physician

Abstract When compassion and the Hippocratic Oath pull Dr. Smith in different directions, Dr. Smith finds a compromise. Dr. Smith acts as the arbiter in bargaining. Knowing this, a patient can take any of three steps to influence the physician's decision: (1) reduce dependency on the physician by gaining access to palliative care; (2) find a physician who has a lot of compassion for patients; and (3) go with an experienced general practitioner, rather than a specialist.

Keywords Compassion · Hippocratic Oath · Two-headed physician

1 Introduction

Dr. Smith faces a dilemma when the patient in the intensive care unit asks her to let him die. Dr. Smith's compassion says: I feel his suffering and I must help him die. But the Hippocratic Oath says: "…I will apply dietetic measures for the benefit of the sick according to my ability and judgment; I will keep them from harm and injustice. I will neither give a deadly drug to anybody who asked for it, nor will I make a suggestion to this effect…." Dr. Smith has compassion, but also has internalized the oath as a commitment to not letting a patient die

This chapter is largely based on Lee (2008).

L. W. Lee, *Behavioral Economics and Bioethics*,
Palgrave Advances in Behavioral Economics,
https://doi.org/10.1007/978-3-319-89779-0_5

(Veatch 1981, pp. 164–168; SUPPORT 1995, pp. 1591–1592). There is a conflict between two emotions. Dr. Smith must resolve the conflict. So how does Dr. Smith do it?

My story is that Dr. Smith balances these emotions by working out a compromise that is fair and just. The compromise is encapsulated in the Nash Solution, a common plot in stories about bargaining. It is a simple and intuitive solution to a complex problem of bioethics.

2 The Patient

Imagine a patient who wishes to die so as to be relieved of pains and inconveniences of living. He does not want to have feeding tubes if he becomes unconscious, or receive electric shocks when his heart stops beating.[1] The patient is old and sick.[2] He also seems totally sincere and fully conscious.[3]

To get a handle on what the patient is thinking, let's say that he feels a level of well-being at present and at any moment in the future. Call that well-being his "total utility." The utility may grow even after the terminal illness has set in. The moment will come, however, when the utility begins a continuous decline. Figure 1 illustrates the utility by curve U. Shortly before his utility begins to decline, the patient may experience stretches of good days and bad days (i.e., bumps not visible on the U curve) but overall he still enjoys living (i.e., the U curve lies above the horizontal axis). Eventually, his health deteriorates and suffering intensifies to such a degree that his utility U begins to decline precipitously.[4] At time t', his utility becomes zero: he loses the desire to live.

As to how the patient develops the utility curve U, I suppose that he depends on prognosis by the physician. Prognosis bookmarks the decline of health in terms of specific events such as "dementia" and "ventilator." The patient does not understand these events, so he builds the physician's prognosis into his forecast of total utility.

[1] See O'Brien et al. (1995), and Fried et al. (2002).

[2] See Patrick et al. (1997), Hammes and Rooney (1998), Covinsky et al. (2000), Fried et al. (2002), and Rietjens et al. (2005).

[3] See Danis et al. (1994), Hammes and Rooney (1998), Carmel and Mutran (1999), and Ditto et al. (2003).

[4] Hamermesh and Soss (1974) explain suicide by a declining utility.

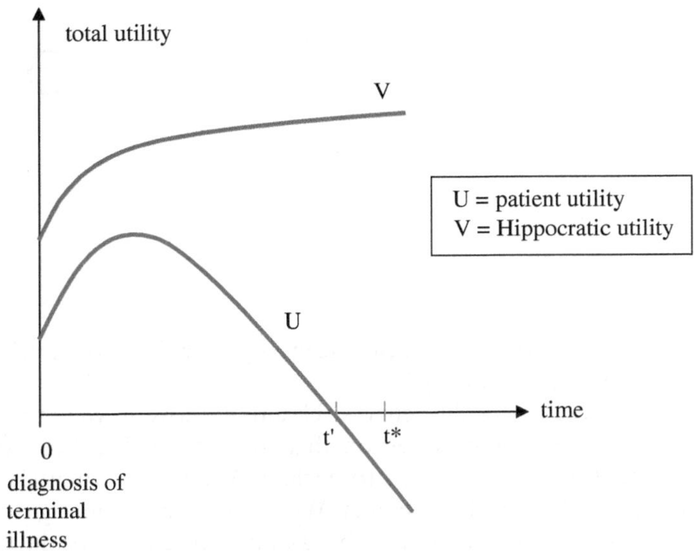

Fig. 1 End-of-life utility functions

3 The Physician

Dr. Smith feels obligated to make good on the Hippocratic Oath. Her feeling of duty can be expressed by a "Hippocratic utility", which is curve V in Fig. 1. The Hippocratic utility rises continuously, reflecting her commitment to saving a life.

At the same time, Dr. Smith feels compassionate toward the dying patient. Standard College Dictionary defines compassion as "pity for the suffering or distress of another, with the desire to help or spare." American Heritage Dictionary defines compassion as "the deep feeling of sharing the suffering of another in the inclination to give aid or support, or to show mercy." In both definitions, compassion includes a desire to help relieve another person's misery. Thus a compassionate physician is not just one who feels pity or sympathy for the suffering patient but one who wants to help the patient put an end to the suffering.

So the burden of making a decision falls on Dr. Smith. She finds herself in the position to determine the manner as well as the timing of the

Fig. 2 The two-headed Dr. Smith

patient's death. She thinks with two heads, one for the patient utility and one for the Hippocratic utility.[5]

In Fig. 1, these two utility curves, U and V, move in the same direction initially, so there is no question that the patient shall live. At some point in time, however, U begins to decline. And that is when Dr. Smith begins to feel the tug of the dilemma: When to let the patient go?

Dr. Smith does not feel comfortable about having conversations about how to end the patient's life. Neither does the patient welcome such conversations. A study of thousands of terminally ill patients found that they do not talk to their physicians much. In spite of—perhaps because of—the gravity of the issue, neither the patient nor the physician makes a serious effort to communicate explicitly with the other (SUPPORT 1995, p. 1591). Physicians ignore patients' wishes, despite mediation by nurses specially trained to improve communication (Fig. 2).

Sooner or later, Dr. Smith plays the role of the arbiter in a game of bargaining. As the arbiter, she uses a "moral compass" to make life-and-death decisions. Her patients can sense that compass in her and they trust her with that role. This explains why most dying patients are passive and why little negotiation, in fact, takes place between physicians and patients in end-of-life decisions.

A popular model of the moral compass is the Nash Solution (Luce and Raiffa 1957). It is simple, and it embodies a form of justice. This solution involves finding the compromise that produces the most net total benefit

[5]This explains why the physician does not have the incentive to be totally forthright with the patient about prognosis. Christakis (1999) has observed that physicians, when confronting dying patients, do not like to prognosticate. This is consistent with the finding that, when provided prognostic information from a state-of-the-art statistical model, most physicians simply ignore it (SUPPORT 1995).

to the parties and then dividing it equally between them. The net benefit to a party is measured against the party's "bargaining position." For the patient, that depends on how much utility the patient expects to have without Dr. Smith's assistance. The patient has a strong bargaining position if, for example, he is receptive to alternative, palliative medicine or is capable of ending own life without assistance. Otherwise, the patient would be more dependent on Dr. Smith and therefore has a weak bargaining position. Dr. Smith's bargaining position as the Hippocratic physician, on the other hand, is her Hippocratic utility if the patient refuses her help. A compassionate physician like Dr. Smith is bound by both her compassion for the patient and the Hippocratic Oath.

As Fig. 1 shows, Dr. Smith finds the Nash Solution at time t^*. It is a later time than the patient's wish t', but it is not infinity either.

4 Lessons for the Patient

If the Nash Solution is how Dr. Smith makes her decision and if the patient knows it, then what steps can the patient take to influence it? Three steps come to mind:

1. Reduce dependency on Dr. Smith by building a strong bargaining position. For example, sign a no-resuscitation order, write a detailed living will, appoint a healthcare proxy, and make an arrangement for palliative care and hospice.
2. Make sure Dr. Smith has more compassion for patients than the other doctors. (This is hard to do.)
3. Make sure Dr. Smith is an experienced general practitioner, rather than a specialist, who tends to feel especially duty-bound by the Hippocratic Oath.

References

Carmel, Sarah, and Elizabeth Mutran. "Stability of Elderly Persons' Expressed Preferences Regarding the Use of Life-Sustaining Treatments." *Social Science and Medicine*, 49 (3), August 1999, pp. 303–311.

Christakis, Nicholas A. *Death Foretold: Prophecy and Prognosis in Medical Care.* Chicago: University of Chicago Press, 1999.

Covinsky, Kenneth E., et al. "Communication and Decision-Making in Seriously Ill Patients: Findings of the Support Project." *Journal of the American Geriatrics Society*, 48 (5), Supplement (May 2000), pp. S187–S193.

Danis, M., J. Garrett, R. Harris, and D. L. Patrick. "Stability of Choices About Life-Sustaining Treatments." *Annals of Internal Medicine*, 120 (7), 1 April 1994, pp. 567–573.

Ditto, Peter H., et al. "Stability of Older Adults' Preferences for Life-Sustaining Medical Treatment." *Health Psychology*, 22 (6), November 2003, pp. 605–615.

Fried, Terri R., Elizabeth H. Bradley, Virginia R. Towle, and Heather Allore. "Understanding the Treatment Preferences of Seriously Ill Patients." *New England Journal of Medicine*, 346 (14), 4 April 2002, pp. 1061–1066.

Hamermesh, Daniel S., and Neal M. Soss. "An Economic Theory of Suicide." *Journal of Political Economy*, 82 (1), January/February 1974, pp. 83–98.

Hammes, B. J., and B. L. Rooney. "Death and End-of-Life Planning in One Midwestern Community." *Archives of Internal Medicine*, 158 (4), 1998, 383–390.

Lee, Li Way. "Compassion and the Hippocratic Oath." *Journal of Socio-Economics*, 37 (5), October 2008, pp. 1724–1728.

Luce, R. Duncan, and Howard Raiffa. *Games and Decisions*. New York: Wiley, 1957.

O'Brien, L. A., J. A. Grisso, G. Maislin, K. LaPann, K. P. Krotki, P. J. Greco, E. A. Siegert, and L. K. Evans. "Nursing Home Residents' Preferences for Life-Sustaining Treatments." *Journal of the American Medical Association*, 274 (22), 1995, pp. 1775–1779.

Patrick, Donald L., Robert A. Pearlman, Helene E. Starks, Kevin C. Cain, William G. Cole, and Richard F. Uhlmann. "Validation of Preferences for Life-Sustaining Treatment: Implications for Advance Care Planning." *Annals of Internal Medicine*, 127 (7), 1997, pp. 509–517.

Rietjens, Judith, Agnes van der Heide, Elsbeth Voogt, Bregie D. Onweteaka-Philipsen, Paul J. van der Maas, and Gerrit van der Wal. "Striving for Quality or Length at the End-of-Life: Attitudes of the Dutch General Public." *Patient Education and Counseling*, 59 (2), November 2005, pp. 158–163.

SUPPORT Principal Investigators. "A Controlled Trial to Improve Care for Seriously Ill Hospitalized Patients: The Study to Understand Prognoses and Preferences for Outcomes and Risks of Treatments." *Journal of the American Medical Association*, 274 (20), 22/29 November 1995, pp. 1591–1598.

Veatch, Robert. M. *A Theory of Medical Ethics*. New York: Basic Books, 1981.

Check for updates

CHAPTER 6

The Governance of Death

Abstract The living will and the health-care agent together govern one's death. As death becomes more complex, the governance structure also undergoes change. First, the agent plays a relatively more active role. Second, the governance draws in physicians, nurses, and ethicists. Death becomes increasingly a compromise between "my wish" and "their wish." I don't see anything wrong with that.

Keywords Death · Living will · Health-care agent · Governance

1 Introduction

I can imagine the last few days of my life when I lie on a hospital bed. Physicians and nurses will be buzzing around me.[1] I know they have come to preserve my life. As a behavioral economist, I also know that they would not be able to prolong my life to the biological maximum even if each of them should want to. You see, it would take

[1]Old persons in general are more likely to die at hospitals (Lynn and Adamson 2003, pp. 1–2). Old persons with dementia are more likely to die at nursing homes (Mitchell et al. 2005).

This chapter is based on Lee (2009).

L. W. Lee, *Behavioral Economics and Bioethics*,
Palgrave Advances in Behavioral Economics,
https://doi.org/10.1007/978-3-319-89779-0_6

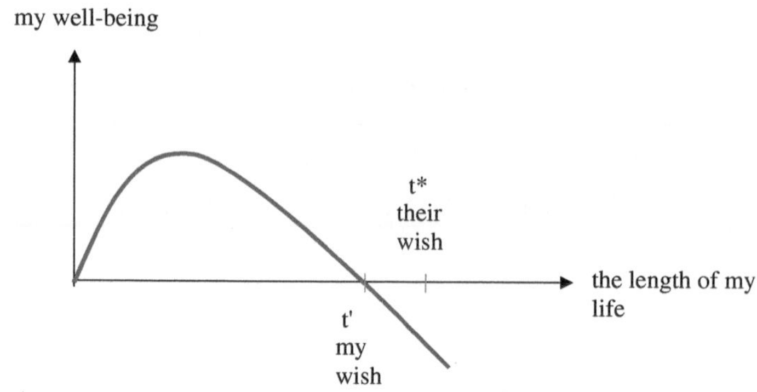

Fig. 1 My well-being as a function of the length of my life

perfect coordination among them to do that. In the real world, perfect coordination is extremely rare. I am not even sure that the physicians and the nurses coordinate at all. For example, they may simply react to each other: The physician would determine how much effort to make by observing how much effort the nurse is making, and vice versa. They end up not maximizing or minimizing anything.[2]

There are two wishes here. "Their wish" is whatever results from the nurse–physician co-operation. "My wish" is what would maximize my well-being as I see it. Figure 1 illustrates the two wishes. Note that my wish is a shorter life than their wish. That is an assumption based on my observation, my reading of others' reports, and my own experience.

So which wish will prevail? If neither will, how are they compromised? That is the question I ask in this chapter.

2 The Governance of Death

How do I express my wish and corral the physicians and nurses around it? Today, I can write a "living will" and appoint someone as my "health-care agent." (Together they form my "advance directive.")

[2]Hofling et al. (1966) and Chambliss (1996) and numerous case studies suggest that physicians and nurses in general hospitals are only weakly bonded, their encounters brief and their communications imperfect. Contrary to conventional wisdom, nurses make critical diagnostic decisions independently of physicians, exercising a large degree of discretion (Jacobson et al. 1998/1999).

They are meant to work in tandem, governing my death in much the same way bylaws and managerial staff govern a corporation. Bylaws specify decisions in situations that are anticipated, while the managerial staff makes decisions in situations that are not anticipated.

2.1 Living Will and Its Limits

My living will is all about nonintervention: a set of multiple choices of medical procedures to be excluded, such as cardiopulmonary resuscitation, mechanical breathing, intravenous feeding, antibiotics, and blood transfusion.

There are good reasons for the narrow scope of my instructions. First, I know little about the organization of medicine in a hospital and the science of medicine itself: I cannot predict exactly what the physicians and the nurses will do to me and what the consequences of their actions are. Also, I do not know how to specify fully the technology to be or not to be applied, or how it is to be combined with other factors like nursing and hospice.

Today, the typical template for living will—find it in Google—emphasizes technology. I think the reason is that technology is easier to specify than labor. It is easier to imagine what a "ventilator" does to me, than what a nurse does to me. In any case, the typical living will is what lawyers would call an "incomplete contract."

There are other limits to a living will. Advances in technology continue to extend the biological life, not only the period in which one is healthy, but also the period of disability before death where one is frail and sick. One result is that by the time the average person dies, he is likely to have suffered several years of multiple physical and mental problems, or what is known as "multifactorial frailty" (Lynn and Adamson 2003). For example, it will be more common for an old patient to be suffering from bone fractures, dementia, incontinence, congestive heart failure, kidney failure, and cancers—all at the same time. These "comorbidities" call for more sophisticated medical decisions than a living will can possibly supply. Does one really want to check the "Do Not Resuscitate" box on the living will if one's heart could stop beating unexpectedly because of a medication that one is taking for high blood pressure? Most people would either not check the box or devise a more complex living will.[3]

A complex living will, however, would be self-defeating. For one thing, physicians do not spend a lot of their time deciphering living wills. Living wills must be read and interpreted, and that takes time and effort. A twenty-page document composed in the manner of a business contract is asking to be misinterpreted or ignored. The important reader of one's living will (i.e., a physician or a nurse) has only seconds or minutes to make a resuscitation decision, or risk letting the patient die prematurely. Indeed, the limits to living wills are precisely the same as the limits to the price system, as explained by Ronald Coase (1937). There are costs of drafting, interpreting, and enforcing contracts. These transaction costs rise with the complexity of a contract.

Moreover, there is institutional resistance. Living wills are fundamentally antithetical to the Hippocratic Oath, which says that everything possible must be done to preserve a life. According to a study of seriously-ill hospital patients (SUPPORT 1995), medical staff do not pay much attention to what patients want or do not want. The majority of physicians in the study (53%) had no idea about their patients' wishes regarding cardiopulmonary resuscitation. The problem seems deep-rooted. The study found that several measures designed to help physicians communicate better with patients did not make a difference. Those measures include facilitation of communication by special nurses,

[3]The President's Council on Bioethics is a public critic of living wills. In its 2005 report, it argues that living wills are immoral. In their view, a living will is an instrument of oppression by a "young self" against an "old self." The old self did not write or have any say about the living will; furthermore, the old self, being disabled by dementia or coma, is not in a position to disagree with the young self. Even if the young self were not duplicitous and meant no harm in drafting the living will, the young self could not have foreseen and understood the complex circumstances and preferences of the old self. So living wills cannot be trusted. This is an imaginative application of the multiple-self models that economists have also used.

However, anyone who distrusts living wills ought to distrust health-care agency as well. A patient's agent, after all, is not the unconscious patient. The agent has no more understanding of the patient's wish than the patient when he was healthy and wrote the living will. The agent has no more authenticity than the patient's healthier and younger self. So I am not at all swayed by the argument of the President's Council on Bioethics against living wills.

better prognostic information to physicians, and better documentation of patient and family preferences.

Finally, the typical patient with a living will does not understand what it is all about. Thorevska et al. (2005) discuss living wills with hospital patients admitted for acute life-threatening problems. They conclude that of the 82 patients with living wills most did not really understand them. For example, only 19% of them knew the prognosis after cardiopulmonary resuscitation. Many also made unwarranted assumptions about the conditions under which the living will would be applied or not applied. The researchers attribute the poor understanding to the fact that most of the living wills were drafted with the aid of lawyers and family members (76%) rather than physicians (7%).

2.2 Agency and Its Limits

A substitute for my living will is my health-care agent—the person that I designate as my surrogate when it comes to determining the manner of my death.[4] The agent serves as my surrogate when I become demented or comatose. My agent is authorized to enforce my living will and, above all, make decisions not covered in my living will.

Agency has its own limitations as a tool of governance. The most serious is that I can't buy it in the market as I can buy apples and haircuts. Below, verbatim, are the seven "qualities of ideal proxies" identified by Terman (2007, p. 224; all emphases his):

1. "They care about you and about honoring your end-of-life wishes;
2. They will be **able** and **available** when you cannot speak for yourself;
3. They are willing to learn about your end-of-life values and treatment preferences from you and your physician;
4. They can be **trusted** to set aside **their** values and preferences as they make decisions based (on) knowing your values and preferences;
5. They are **assertive**, **articulate**, and **responsive**;

[4] The agent is variously known as "health care proxy," "proxy," "surrogate," or "power of attorney." The agent is a person, not a document.

6. They have an adequate **knowledge** of the challenges and workings of the health care and legal systems, OR are sufficiently **resourceful** to learn what they need to know; for example, they can ask an ethics committee or an elder-care or estate attorney to consult; and,
7. They have sufficient **energy**, **persistence**, and **diplomacy** to vehemently advocate your Last Wishes."

Because ideal proxies are hard to find, one is likely to have to settle for an agent with just one or two of these qualifications. Some people simply cannot find satisfactory proxies, and they can become despondent. In 2004, Maryland Supreme Court Judge Robert Hammerman, who had early signs of Alzheimer's, committed suicide. On the day before he shot himself, he sent a ten-page letter to 2000 friends and acquaintances. He said, "There are happily certain people who care about me – but none able to care for me" (Terman 2007, pp. 324–325). He dreaded living in a nursing home; however, he could have easily made arrangement to be cared for at home by visiting physicians and nurses. He did not because he was afraid of losing his mind to Alzheimer's. And he was afraid of losing his mind because he could not find among the 2000 people anyone whom he could trust as his agent when he could no longer speak for himself.

3 The Future of Governance

The human population continues to become older. So what does the future hold for the governance of death? Here I think of a lesson that I have learned from the literature on economic organizations (Williamson 1995; Lafontaine and Slade 2007):

> As the complexity and the uncertainty facing an organization grow, the organization will rely more on administrative decision making instead of explicit contracting.

That is, as we die older and the technological and institutional complexities surrounding our death grow, we depend more on our agents to govern death. More complexities bring about more occasions when our agents will make on-the-spot decisions and act aggressively to protect us from other people's wishes. These occasions will not be covered in the living wills.

Fig. 2 The expansion path of governance

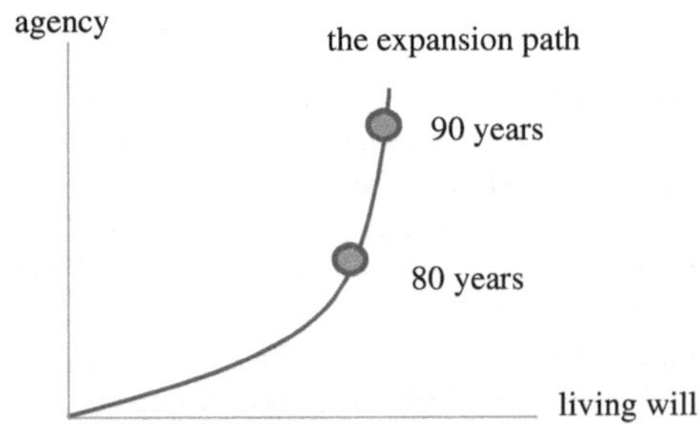

Figure 2 illustrates the lesson learned. The governance structure at any age is composed of doses of living will and agency. As age advances, there are greater doses of both; proportionally, however, there will be more agency. That is why the path becomes steeper.

When looking at this graph, I see a growing demand for agency. Today, most states have passed legislation to recognize health-care proxies. Specialists like elder-law attorneys, professional ethicists, and end-of-life consultants are emerging.[5] They are already working side by side with all the other people one must trust in one's life: physicians, dentists, accountants, and barbers.

4 Conclusion

In this chapter, I look at how people come together to deal with a person's death. I learn most from stories about hospital wards, the ICUs, and the nursing homes. These are stories by people with their boots on the ground: nurses, physicians, patients' agents, patients, and patients' family members (American Society for Bioethics and Humanities 2017; Gawande 2014; Zitter 2017).

By these accounts, one's living will and health-care agent together represent how one wishes to die. As death becomes more complex, two things happen. First, the agent plays a relatively greater role. Second, the governance increasingly draws in physicians, nurses, and ethicists. Death becomes increasingly a compromise between "my wish" and "their wish." I see justice in that.

[5] For example, American Society for Bioethics and Humanities (www.asbh.org).

References

American Society for Bioethics and Humanities. *Addressing Patient-Centered Ethical Issues in Health Care: A Case-Based Study Guide*. 2017. www.asbh.org.

Chambliss, Daniel. *Beyond Caring: Hospitals, Nurses and the Social Organization of Ethics*. Chicago: The University of Chicago Press, 1996.

Coase, Ronald. "The Nature of the Firm." *Economica*, 4, November 1937, pp. 386–405.

Gawande, Atul. *Being Mortal: Medicine and What Matters in the End*. Farmington Hills, MI: Gale Cengage Learning, 2014.

Hofling, Charles K., Eveline Brotzman, Sarah Dalrymple, Nancy Graves, and Chester M. Pierce. "An Experimental Study in Nurse-Physician Relations." *The Journal of Nervous and Mental Disease*, 143 (2), 1966, pp. 171–180.

Jacobson, Peter D., Louise E. Parker, and Ian D. Coulter. "Nurse Practitioners and Physician Assistants as Primary Care Providers in Institutional Settings." *Inquiry*, 35 (4), 1998/1999, pp. 432–446.

Lafontaine, Francine, and Margaret Slade. "Vertical Integration and Firm Boundaries: The Evidence." *Journal of Economic Literature*, 45 (3), 2007, pp. 629–685.

Lee, Li Way. "Living Will: Ruminations of an Economist." *Journal of Socio-Economics*, 38 (1), January 2009, pp. 25–30.

Lynn, Joanne, and David M. Adamson. "Living Well at the End of Life: Adapting Health Care to Serious Chronic Illness." Rand Health White Paper, 2003, WP-137.

Mitchell, Susan L., Joan M. Teno, Susan C. Miller, and Vincent Mor. "A National Study of the Location of Death for Older Persons with Dementia." *Journal of the American Geriatrics Society*, 53 (2), 2005, pp. 299–305.

SUPPORT Principal Investigators. "A Controlled Trial to Improve Care for Seriously Ill Hospitalized Patients: The Study to Understand Prognoses and Preferences for Outcomes and Risks of Treatments." *Journal of the American Medical Association*, 274 (20), 1995, pp. 1591–1598.

Terman, Stanley A. *The Best Way to Say Goodbye*. Carlsbad: Life Transitions Publications, 2007.

The President's Council on Bioethics. *Taking Care: Ethical Caregiving in Our Aging Society*. Washington, DC, September 2005. www.bioethics.gov

Thorevska, Natalya, Lisa Tilluckdharry, Sumit Tickoo, Andrea Havasi, Yaw Amoateng-Adjepong, and Constantine Manthous. "Patients' Understanding of Advance Directives and Cardiopulmonary Resuscitation." *Journal of Critical Care*, 20 (1), 2005, pp. 26–34.

Williamson, Oliver E. *The Mechanisms of Governance*. Oxford: Oxford University Press, 1995.

Zitter, Jessica. *Extreme Measures: Finding a Better Path to the End of Life*. New York, Avery: An Imprint of Penguin Random House, 2017.

PART III

Young People and Old People

CHAPTER 7

The Public Health Roulette

Abstract Changes in life expectancies perturb the balance of justice between the young and the old, prompting reallocation of income and health. The long-term consequences are difficult to predict. I report a case where greater life expectancies create greater health disparity between the generations.

Keywords Life expectancy · Public health · Intergenerational conflict
Health disparity

1 Introduction

Our life expectancies have been going up rapidly. That's an excuse for drinking a lot of champagne. But, on second thought, it is also a cause for ulcer. Life expectancy is nothing without "health expectancy." To see my health expectancy, I take my total lifetime income and divide it by my life expectancy. The average annual income is a good predictor of health (Ettner 1996; Lynch et al. 2004; Wagstaff and van Doorslaer 2000). That is, the following equation holds pretty well (Fig. 1).

By this equation, my health expectancy can go down even as my life expectancy goes up. A longer life is not necessarily a healthier (and happier) life!

L. W. Lee, *Behavioral Economics and Bioethics*,
Palgrave Advances in Behavioral Economics,
https://doi.org/10.1007/978-3-319-89779-0_7

Fig. 1 The health equation

$$\text{Health} = \frac{\text{Total Income}}{\text{Life Expectancy}}$$

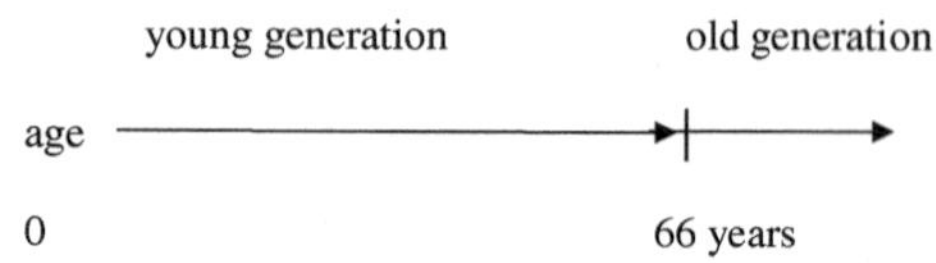

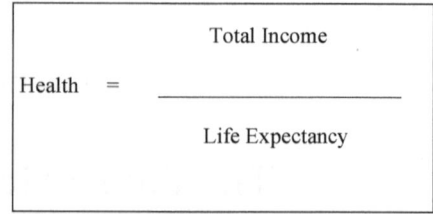

Fig. 2 Two generations

In this chapter, I visit a population in search of this anomaly. The first thing that catches my eye is the Social Security Program. This program divides the population into two generations: young and old. Anyone younger than 66 years belongs to the young generation, while anyone 66 or older belongs to the old generation. If people in the population line up by age, we will see two groups (Fig. 2).

The Social Security Program requires the young generation to send a portion of their income to the old. Since health depends on income, the Social Security Program practically requires the young to send a portion of their health to the old.

The Social Security Program is a social contract. It is built on economics and ethics. Richard Posner (1995, pp. 84–94) tells a tale of the struggles beneath the program. He employs the metaphor of a person embodying two selves.[1] The two selves have profoundly different outlooks on life, so that they are strangers to each other. The old self, who lives in the future, can make its presence felt today through surrogacy: "... the current old, in effect as proxies for the future old selves of the current young, struggle with the current young in the political marketplace for the allocation of consumption over the life cycle" (1995, p. 267). The metaphor is consistent with what we know about young people: they are impatient (Read and Read 2004), and they don't like to

[1] Behavioral economists in particular have been using two-generation models in their work on self-command (Schelling 1984), mental accounting (Thaler 1985), generation control (Thaler and Shefrin 1981), and hyperbolic discounting (Laibson 1997).

save for old age (Sunstein and Thaler 2003; Thaler and Benartzi 2004; Akerlof and Shiller 2009, Chapter 10).

2 The Social Security Program: An Example

Let's say the population consists of Junior and Senior. They bargain about the distribution of income.[2] Their agreement—the Social Security Program—is the Nash Solution to bargaining (Luce and Raiffa 1957, pp. 126–128).

The Nash Solution depends on Junior's and Senior's "bargaining positions." Junior has a life expectancy of 60 years and earns a total of $2 million over that period. Junior must choose between two options: share that income with Senior tax-free (by way of the Social Security Program), or pay 40% tax on it (to the Internal Revenue Service). Senior, who has a life expectancy of 20 years, must choose between taking a share of Junior's income (via the Social Security Program) or live on $0.2 million (via charity and Medicaid.)

With these bargaining positions, the Nash Solution is that Junior and Senior share the $2.0 million three-to-one: $1.5 million for Junior and $0.5 million for Senior. Given their life expectancies of 60 years and 20 years, have you noticed that the Nash Solution results in total equality in annual incomes: $25,000 a year each? That also means equality in health, inasmuch as annual income correlates with health.[3]

I see justice in the Social Security Program.

[2]Schelling (1984, Chapters 3 and 4) sees the old and the young generations trying to resolve conflict through bilateral bargaining. The conflict between the two selves is mediated by the social context in terms of conventions, ethics, laws, and entitlement programs, which penalize a failure to bargain in good faith on the part of any one generation.

[3]Here are some accounting details. First, Junior will transfer to Senior a total of $500,000 for retirement ($25,000 × 20), implying a saving rate of 25% ($500,000/$2,000,000). Second, if Junior does not save for retirement, then he pays 40% of the total income, or $800,000, in tax. In this case, the society siphons more than enough of the $2 million to cover Social Security payment to Senior, which is $10,000 a year for 20 years, or $200,000. Indeed the society would end up with a surplus of $600,000 as a result of Junior's imprudence.

A Special Note: If Junior and Senior bargain to share *both* $2 million *and* 80 years of life expectancy, the terms of the Nash Solution will be exactly those in the "example" in the text: Junior would live for 60 years on $1.5 million; Senior would live for 20 years on $0.5 million. The proof is surprisingly tedious.

3 Public Health Agency

As I look around a bit more in this population, another public policy-maker catches my eye: Public Health Agency. The Agency has been successful in increasing life expectancies. Much of the success has resulted from lower rates of mortality. Many people laud that accomplishment. But as the Agency begins to move into the realm of extending (as opposed to saving) lives, some begin to ask questions. For example, Kinsella (2009, p. 22) asks: "Are we living healthier as well as longer lives, or are our additional years spent in poor health?"

Let's take this question seriously. Consider the example of a population with two people: Junior and Senior. Suppose that the Public Health Agency has found a way to extend either Junior's life by 10 years, or Senior's life by 10 years, but not both. What happens to the distribution of income between them?

The answer depends on whose life is extended. Let's consider two cases.

Case 1 Suppose the Public Health Agency discovers a drug that reduces the probability of deadly measles infecting young people. With that drug, Junior now expects to live longer by 10 years. This development leads to renegotiation between Junior and Senior about income distribution. Before the drug, Junior and Senior each spent $25,000 a year. (Recall the Example above.) Now, after renegotiation, Junior keeps $1.6 million, giving Senior $0.4 million. That is, Junior now spends $22,857 (for the longer expected period of 70 years) while Senior spends $20,000 a year (for the same expected period of 20 years). So annual incomes of both generations will drop from the existing level of $25,000: by $2143 for Junior and by $5000 for Senior. As a result of the anti-measles program, therefore, health of the population declines across the board, with Senior suffering more than Junior.

Case 2 Suppose that the Public Health Agency finds a drug that totally eradicates Parkinson's and Alzheimer's, thereby increasing Senor's life expectancy by 10 years. (The drug does not do anything for Junior.) Again, this drug will affect the Social Security program, by triggering a renegotiation of the division of the $2 million total income between Junior and Senior. I have found that, under the new terms, Junior will

Table 1 Health disparity

Life expectancies (Junior, Senior)	*Junior health (annual income in $1000s)*	*Senior health (annual income in $1000s)*	*Disparity index (Junior health/ Senior health)*
(60, 20)	25	25	25/25 = 1.00
(70, 20)	23	20	23/20 = 1.15
(60, 30)	24	18	24/18 = 1.33

keep $1.45 million, transferring the remaining $0.55 million to Senior. That is, Junior will have annual income of $24,167 and Senior will have $18,333 a year. These are less than the $25,000 annual income before the program. Given the income-health nexus, these represent declines in health. Note that Senior loses a lot more than Junior, creating disparity in income as well as health.

These two cases are summarized in Table 1. I find most striking the progression of the disparity index. It suggests that the old generation holds the shorter end of "the health stick," no matter whose life is extended.

4 The Public Health Feedback Loop

In this population, an increase in life expectancy will trigger income redistribution. Intergenerational conflict will ensue, threatening the cohesion of democracy (Kaplan et al. 2002; Sokolovsky 2009). So the Public Health Agency can destabilize the Social Security Program.

There is a feedback loop here. Public Health Agency blesses me with a longer old life, which sets off income redistribution, which sets off health redistribution, which in turn sets off life-expectancy redistribution, which will spur Public Health Agency into some more actions. This feedback loop, as Fig. 3 suggests, is the result of the breaking down of the equality in Fig. 1:

Intergenerational justice at any moment, therefore, is fragile: Attempts to change it at any moment will create uncertainty. Large, heroic attempts will interact with other issues of life and death to plunge us into greater depths of the unknown and the unknowable (Kotlikoff and Burns 2014).

Fig. 3 The health feedback loop

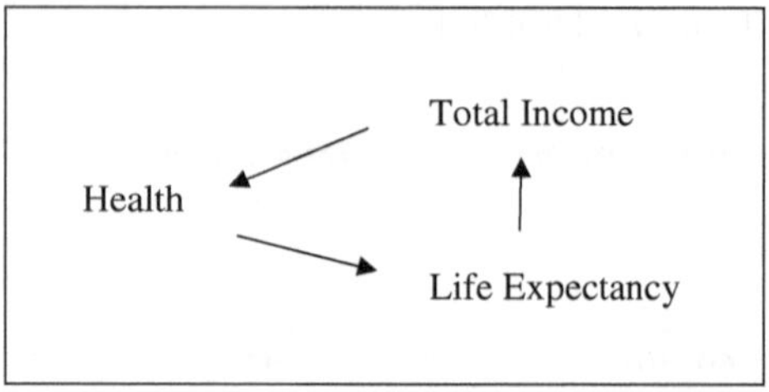

5 The Principle of Minimum Health Disparity

At the end of my visit, I am left with a nagging question: What ought to be the goal of the Public Health Agency? It looks as if the Agency is playing roulette with its policies of life extension, however well-meaning they may seem at the moment.

A reasonable goal of Public Health Agency, I think, is "Minimum Health Disparity." This would require the Agency to choose the policy that entails the least health disparity in the long run, subject to an overall goal of extending the life expectancy of the total population.

Consider an example. Suppose that the Agency wants to add 10 years to the life expectancy of the total population. Suppose further that the agency has only two choices:

1. Raise the life expectancy of the young generation by 10 years;
2. Raise the life expectancy of the old generation by 10 years.

These are precisely the two cases we considered earlier and summarized in Table 1.

In both cases, health disparity goes up. But in the first case, health disparity goes up less. By the Principle of Minimum Health Disparity, the Agency should give the 10 years to Junior.

Appendix

The Nash Solution with a Fixed Total Resource to Share

Junior and Senior bargain for the biggest share of a resource. The Nash solution to the bargaining problem is the solution to the following mathematical program:

$$MAX\left(\frac{R_y}{T_y}-c_y\right)\left(\frac{R_o}{T_o}-c_o\right) \tag{1}$$

with respect to R_y and R_o, subject to a resource constraint $R = R_y + R_o$, where

R_y resource for Junior (the young generation) if bargaining succeeds
R_o resource for Senior (the old generation) if bargaining succeeds
T_y Junior's life expectancy
T_o Senior's life expectancy
c_y Junior's resource if bargaining fails
c_o Senior's resource if bargaining fails

In the Nash Solution, Junior's and Senior's annual incomes are, respectively:

$$\begin{aligned}\frac{R_y}{T_y} &= \frac{R + c_yT_y - c_oT_o}{2T_y}\\ \frac{R_o}{T_o} &= \frac{R - c_yT_y + c_oT_o}{2T_o}\end{aligned} \tag{2}$$

Case 1 How an increase in Junior's life expectancy affects income distribution.

$$\begin{aligned}\frac{\partial\left(\frac{R_y}{T_y}\right)}{\partial T_y} &= -\frac{\left(\frac{R_y}{T_y}-\frac{c_y}{2}\right)}{T_y} < 0\\ \frac{\partial\left(\frac{R_o}{T_o}\right)}{\partial T_y} &= -\frac{c_y}{2T_o} < 0\end{aligned} \tag{3}$$

Case 2 How an increase in Senior's life expectancy affects income distribution.

$$\begin{aligned}\frac{\partial\left(\frac{R_y}{T_y}\right)}{\partial T_o} &= -\frac{c_o}{2T_y} < 0\\ \frac{\partial\left(\frac{R_o}{T_o}\right)}{\partial T_o} &= -\frac{\left(\frac{R_o}{T_o}-\frac{c_o}{2}\right)}{T_o} < 0\end{aligned} \tag{4}$$

References

Akerlof, George A., and Robert J. Shiller. *Animal Spirits*. Princeton: Princeton University Press, 2009.

Ettner, Susan. "New Evidence on the Relationship Between Income and Health." *Journal of Health Economics*, 15 (1), February 1996, pp. 67–85.

Kaplan, Matthew, Nancy Henkin, and Atsuko Kusano. *Linking Lifetimes: A Global View of Intergenerational Exchange*. Lanham: University Press of America, 2002.

Kinsella, Kevin. "Global Perspectives on the Demography of Aging." In Jay Sokolovsky, ed. *The Cultural Context of Aging: Worldwide Perspectives*, 3rd ed. Santa Barbara, CA: Praeger, 2009, pp. 13–29.

Kotlikoff, Laurence, and Scott Burns. *The Clash of Generations: Saving Ourselves, Our Kids, and Our Economy*. Cambridge: MIT Press, 2014.

Laibson, David. "Golden Eggs and Hyperbolic Discounting." *Quarterly Journal of Economics*, 62, May 1997, pp. 443–477.

Luce, R. Duncan, and Howard Raiffa. *Games and Decisions*. New York: Wiley, 1957.

Lynch, John, et al. "Is Income a Determinant of Population Health? Part 1. A Systematic Review." *Milbank Quarterly*, 82, 2004, pp. 5–99.

Posner, Richard A. *Aging and Old Age*. Chicago: The University of Chicago Press, 1995.

Read, Daniel, and N. L. Read, "Time Discounting over the Lifespan." *Organizational Behavior and Human Decision Processes*, 94, 2004, pp. 22–32.

Schelling, Thomas C. *Choice and Consequence*. Cambridge: Harvard University Press, 1984.

Sokolovsky, Jay. *The Cultural Context of Aging: Worldwide Perspectives*, 3rd ed. Santa Barbara, CA: Praeger, 2009.

Sunstein, Cass, and Richard Thaler. "Libertarian Paternalism." *American Economic Review*, 93, 2003, pp. 175–179.

Thaler, Richard. "Mental Accounting and Consumer Choice." *Management Science*, 4, 1985, pp. 199–214.

Thaler, Richard, and Hersh M. Shefrin. "An Economic Theory of Generation Control." *Journal of Political Economy*, 89 (2), 1981, pp. 392–410.

Thaler, Richard, and Shlomo Benartzi. "Save More Tomorrow™: Using Behavioral Economics to Increase Employee Saving." *Journal of Political Economy*, 112 (S1), 2004, pp. 164–187.

Wagstaff, Adam, and Eddy van Doorslaer. "Income Inequality and Health: What Does the Literature Tell Us?" *Annual Review of Public Health*, 21, 2000, pp. 543–567.

CHAPTER 8

The Long Shadow of Caregiving

Abstract Caregiving seems to be a great bargain for old people, not to mention that it is probably a key to the survival of the human species. In this chapter, I visit a population in which young people give care to old people. I note that caregiving has a long shadow: Grow it a bit today, and we won't walk out of its shadow for a much longer while. For example, a decline of death rate of the old people from 5 to 4% today will result in a rise of the caregiving burden from 0.4 to 0.5 old persons per young person. This climb will take 126 years.

Keywords Caregiver · Caregiving · Demography

1 Introduction

Young people today transfer more income to old people than the other way around. Lee and Mason (2011) find that young people are "net givers" in Germany, Austria, Japan, Slovenia, and Hungary. Eggleston and Fuchs (2012) see this intergenerational transfer as a general feature of the "longevity transition" on planet Earth.

But have you noticed that the young people bear another burden: caregiving? This burden gets less mention than income, but it is the elephant in the room. Imagine for a moment that mom and dad can expect to live longer all of a sudden. You will have to take care of them at

L. W. Lee, *Behavioral Economics and Bioethics*,
Palgrave Advances in Behavioral Economics,
https://doi.org/10.1007/978-3-319-89779-0_8

home by cutting back hours of work. As a result, you have lower income and less time for taking care of your own health. These sacrifices have long-run consequences for you, for sure. As I look over the horizon, I see consequences for the rest of us also. For one thing, I see fewer young people like you, and I see a smaller and older population. So I see growing caregiving burden on the average young person. I see trouble with the balance of intergenerational justice.

In this chapter, I pay visit to a population in order to track down the demographic consequences of a build-up in the caregiving burden.

2 A Population with Caregiving

Caregiving comes from two sources: market or home. In either case, caregiving exacts a toll on the health of the caregivers.[1] Studies have shown that caregiving can cause depression,[2] stress,[3] and heart disease,[4] especially among an old person's spouse and daughters. Not surprisingly, caregivers themselves tend to die sooner.[5] Caregiving also adversely affects young people's health through the income effect. With more income going to the old, the young have less income for themselves. Or when they spend more time on caring for their elderly parents, they have less time to earn income. A lower income has adverse effects on health. For a wide range of income, income and health have been shown to be positively related (Deaton 2006; Ettner 1996; Ecob and Smith 1999; Wagstaff and van Doorslaer 2000; Lynch et al. 2004). When the young generation devotes more time and money to the old generation, the young generation in effect "transmits" health to the old generation.[6]

To see how caregiving affects demography, imagine a new pill that causes a surge in longevity among old people. Immediately, there will be more old people. They require caregiving by young people. The burden

[1] Kim and Knight (2008), Christakis and Iwashyna (2003), Ho et al. (2009).

[2] Burton et al. (2003), Covinsky et al. (2003), Haley et al. (2003), McCusker et al. (2007).

[3] Hubbell and Hubbell (2002), Vitaliano et al. (2003), Gaugler et al. (2007).

[4] Lee et al. (2003).

[5] Schultz and Beach (1999), Christakis and Iwashyna (2003).

[6] This is to be distinguished from the conventional definition of "intergenerational transmission of health," which runs one way from parents to children, not from the young to the elderly (Ahlburg 1998).

of caregiving on young people rises, and fewer of them will survive into old age themselves. In a few decades, that means fewer old people. That, in turn, lessens the burden of caregiving. So begins another cycle.

The demographic feedback loop can work in complex ways, so I study it by simulation.[7] In this section, I report the results of a model based on the following assumptions:

1. Every year, 5% of old people leave the population, due to death and emigration. Meanwhile, newly old people join the population.
2. Every year, births add to the number of young people; premature deaths and aging (turning 65, say) subtract from the number of young people. We assume that these natural processes yield a net replacement rate of zero among young people. That is, on the strength of their own number, young people are just able to replace themselves. This assumption serves to avoid confounding effects of net replacement rate with those of caregiving, which is our emphasis.
3. Young people in this population bear the cost of caregiving. The cost is measured by the loss of young lives due to poor health. Let's say that the loss is proportional to the size of the old generation, or 2% of *old* people to be exact.
4. Lastly, 0.8 million immigrants, all young, join the population every year. This assumption is necessary for preventing the population from vanishing under the growing weight of caregiving due to longevity.

A population with these features will have a steady state, where there are 200 million young people and 80 million old people, for a total of 280 million people. A young person, therefore, supports 0.4 old persons on average. In other words, the dependency ratio equals 0.4. See Appendix I for more details.

[7] I am grateful to Yong-Gook Jung for producing the graphs and for giving me the permission to use them in this chapter. He and I have written a paper that features the same population undergoing both demographic and economic changes (Jung and Lee 2012). The graphs appear in that paper as well.

Table 1 Long-run effects of a rise in longevity on population[a]

	The old generation	*The young generation*	*Total population*	*Dependency ratio*
Before	80	200	280	0.40
After	80	160	240	0.50

[a]Population sizes in millions

Now suppose that a surge in longevity makes it possible to reduce the annual death rate in the old generation from 5 to 4%. In the short run, there will be more old people than before. This sets off changes in composition and size of the population. In time, the population will stabilize, this time at only 240 million, representing a decline of 40 million. Table 1 summarizes the populations before and after longevity. Clearly, the entire decline occurs in the number of young people. As a result, there will be proportionally more old people in this population. The dependency ratio will rise from 0.40 to 0.50.

Figure 1 depicts various adjustment trajectories. The old generation increases rapidly for 62 years, reaches 96.4 million, and then begins a gradual decline. It will take 215 years for the old generation to reach within 10% of the new steady-state level. The young generation declines monotonically after the shock. It will take 188 years for the young generation to reach within 10% of the new steady-state level. As to the total population, it grows for 36 years, reaches 291.5 million, and then it begins a descent. It will reach within 10% of the new steady-state level only after 197 years. The initial growth of the old generation and, especially, the decline of the young generation lift the dependency ratio. The ratio reaches the peak at 0.509 old persons per young person in 126 years after the shock and gradually declines toward the new steady-state level at 0.500 old persons per young person.

In other words, about a hundred years after a longevity shock, a number of changes will become noticeable. First, despite the greater longevity, there will *not* be more old people. Second, there will be *fewer* young people. Consequently, the population will be smaller. Most disturbingly, there will be proportionally fewer young people than old people. That means a heavier caregiving burden on the average young member of the future population.

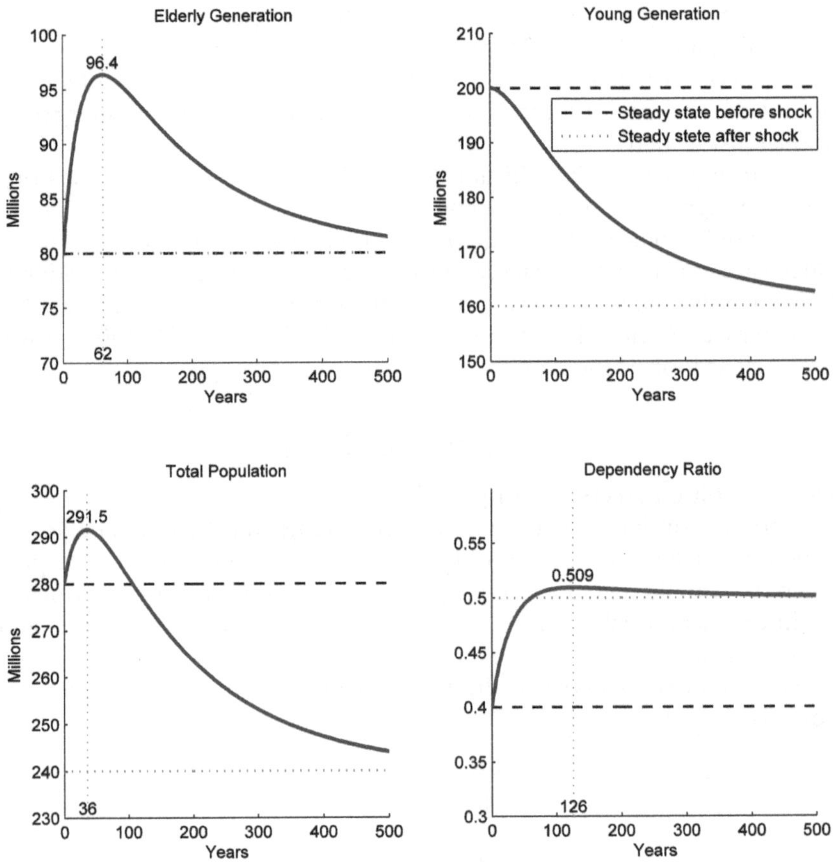

Fig. 1 Trajectories of a population after a longevity shock

3 The Long Shadow of Caregiving

Caregiving seems to be a great bargain for old people, not to mention that it is probably a key to the survival of the human species. In this chapter, I simply notice that caregiving has a long shadow. Grow it a bit today, and we won't walk out of its shadow for a much longer while. More disturbingly, the farther we go into the future, the heavier will be the caregiving burden on young people. In the model that I simulate, a decline of death rate of the old people from 5 to 4% today will result in a rise of the caregiving burden from 0.4 to 0.5 old persons per young person. This climb will take 126 years.

I am concerned that, in our headlong effort to live longer older, we do not see the toll on future generations. There is an issue of intergenerational justice here.

In closing, I note the critical roles that immigration plays in the story. First, immigration is what sustains the population; without it, the population would collapse. Second, young immigrants supply a large share of elder care through the market mechanism. There is empirical evidence that a family caregiver's own health improves with access to the market for elder care (Christakis and Iwashyna 2003; Gaugler et al. 2007). I will take a close look at the cross-border migration of caregivers next.

Appendix I

Demographic Effects of Longevity

The population has two generations, the young and the old. The young generation takes care of the old generation. The amount of care provided to an old person is constant, being determined by a principle of intergenerational justice inherent in the population's culture and institutions.

The interactions between the two generations follow two differential equations:

$$\frac{\delta E}{\delta t} = aY - dE \tag{1}$$

$$\frac{\delta Y}{\delta t} = -cE + I \tag{2}$$

where all variables and coefficients have positive values:

- E the number of old people
- Y the number of young people
- a the rate of aging
- c the burden of caregiving on the young in terms of their lives lost per old person
- d the mortality rate among old people
- I the number of young immigrants who join the population every year

The Old-Generation Equation (1)

The number of old people at any moment changes for two reasons. First, as some young people become old, the number of old grows at a rate proportional to the number of young people. This proportion defines the rate of "ageing." Second, old people die, at a rate proportional to the size of their size. That is the rate of mortality.

The Young-Generation Equation (2)

The number of young people changes for three reasons: caregiving burden, "net replacement" and immigration. First, as explained above, caregiving burden is proportional to the size of the old generation. Second, net replacement rate captures all changes that are proportional to the size of the young generation itself: birth, diseases, accidents, and aging. For example, if in any year 2% of young people give birth, each to a single baby, 0.6% of them die of natural causes, and 2% of them become old, then the net replacement rate equals: b = 0.02 children – 0.006 deaths – 0.02 ageing = –0.006.

Third, the immigration of young people takes place at a steady rate, determined by fixed factors such as immigration policies and migrations from other populations.

To avoid confounding any effects of net replacement rate with those of caregiving, assume that the net replacement rate of the young generation is zero. This assumption, without more, would cause the population to vanish under the weight of caregiving. That is why we assume also that a stead number of young immigrants join the population every year.

Under these assumptions, the two generations have initial steady-state sizes equal to:

$$E = \left(\frac{1}{c}\right)I \tag{3}$$

$$Y = \left(\frac{d}{ac}\right)I \tag{4}$$

Two other characteristics of the population are noteworthy: the total size and the "dependency ratio." The total population is the sum of the old and the young generations:

$$E + Y = \left(\frac{a + d}{ac}\right)I \tag{5}$$

The dependency ratio measures the caregiving burden in the population. It is defined as the average number of old persons supported by a young person:

$$\frac{E}{Y} = \frac{a}{d} \tag{6}$$

The greater the ratio, the greater the caregiving burden on young people.[8] For example, if the dependency ratio is 0.25, then each young person will care for 0.25 of an old person, or four young persons will share the cost of caring for an old person. If the dependency ratio becomes 0.5, then each young will care for one half of an old person, or two young persons will share the cost of caring for one old person. If the ratio is 1.0, then each young person will bear the cost of taking care of one old person. A greater caregiving burden also means more of a young person's income will be transferred to the old generation, with adverse health consequence for the young person.

References

Ahlburg, Dennis. "Intergenerational Transmission of Health." *American Economic Review*, 88 (2), May 1998, pp. 265–270.

Burton, Lynda C., Bozena Zdaniuk, Richard Schulz, Sharon Jackson, and Calvin Hirsch. "Transitions in Spousal Caregiving." *The Gerontologist*, 43, 2003, pp. 230–240.

Christakis, Nicholas A., and Theodore Iwashyna. "The Health Impact of Health Care on Families: A Matched Cohort Study of Hospice Use by Decedents and Mortality Outcomes in Surviving, Widowed Spouses." *Social Science and Medicine*, 57 (3), August 2003, pp. 465–475.

Covinsky, Kenneth E., Robert Newcomer, Patrick Fox, Joan Wood, Laura Sands, Kyle Dane, and Kristine Yaffe. "Patient and Caregiver Characteristics Associated with Depression in Caregivers of Patients with Dementia." *Journal of General Internal Medicine*, 18 (12), December 2003, pp. 1006–1014.

[8]The United Nations (2006, Table A.III.3) reports old-age dependency ratios. An old person is defined as anyone 65 years or older. A young person is anyone with age between 15 and 64 years.

Deaton, Angus. "Global Patterns of Income and Health: Facts, Interpretations, and Policies." Working Paper no. 12735, National Bureau of Economics Research, December 2006.

Ecob, Russell, and George D. Smith. "Income and Health: What is the Nature of the Relationship?" *Social Science and Medicine*, 48 (5), March 1999, pp. 693–705.

Eggleston, Karen N., and Victor R. Fuchs. "The New Demographic Transition: Most Gains in Life Expectancy Now Realized Late in Life." *Journal of Economic Perspectives*, 26 (3), Summer 2012, pp. 137–157.

Ettner, Susan. "New Evidence on the Relationship Between Income and Health." *Journal of Health Economics*, 15 (1), February 1996, pp. 67–85.

Gaugler, Joseph E., Anne M. Pot, and Steven H. Zarit. "Long-Term Adaptation to Institutionalization in Dementia Caregivers." *The Gerontologist*, 47 (6), 2007, pp. 730–740.

Haley, William E., Laurie A. LaMonde, Beth Han, Allison M. Burton, and Ronald Schonwetter. "Predictors of Depression and Life Satisfaction among Spousal Caregivers in Hospice: Application of a Stress Process Model." *Journal of Palliative Medicine*, 6 (2), April 2003, pp. 215–224.

Ho, Suzanne, Alfred Chan, Jean Woo, Portia Chong, and Aprille Sham. "Impact of Caregiving on Health and Quality of Life: A Comparative Population-Based Study of Caregivers for Elderly Persons and Noncaregivers." *The Journals of Gerontological Series A: Biological Sciences and Medical Sciences*, 64A (8), 2009, pp. 873–879.

Hubbell, Larry, and Kelly Hubbell. "The Burnout Risk for Male Caregivers in Providing Care to Spouses Affiliated with Alzheimer's Disease." *Journal of Health Human Service Administration*, 25 (1), Summer 2002, pp. 115–132.

Jung, Yong-Gook, and Li Way Lee, "Longevity Transition and Economic Decline." Draft, December 2012.

Kim, Jung-Hyun, and Bob G. Knight. "Effects of Caregiver Status, Coping Styles, and Social Support on the Physical Health of Korean American Caregivers." *The Gerontologist*, 48, 2008, pp. 287–299.

Lee, Ronald D., and Andrew Mason. "Generational Economics in a Changing World." *Population and Development Review*, 37 (s1), 2011, pp. 115–142.

Lee, Sunmin, Graham A. Colditz, Lisa F. Berkman, and Ichiro Kawachi. "Caregiving and Risk of Coronary Heart Disease in U.S. Women: A Prospective Study." *American Journal of Preventive Medicine*, 24 (2), February 2003, pp. 113–119.

Lynch, John, George D. Smith, Sam Harper, Marianne Hillemeir, Nancy Ross, George A. Kaplan, and Michael Wolfson. "Is Income Inequality a Determinant of Population Health? Part 1: A Systematic Review." *Milbank Quarterly*, 82 (1), March 2004, pp. 5–99.

McCusker, Jane, Eric Latimer, Martin Cole, Antonio Ciampi, and Maida Sewitch. "Major Depression Among Medically Ill Elders Contributes to Sustained Poor Mental Health in Their Informal Caregivers." *Age and Ageing*, 36 (4), 2007, pp. 400–406.

Schultz, Richard, and Scott Beach. "Caregiving as a Risk Factor for Mortality." *Journal of the American Medical Association*, 282, 1999, pp. 2215–2219.

United Nations. *World Population Ageing 2007.* Population Division, 2006.

Vitaliano, Peter P., Jianping Zhang, and James M. Scanlan. "Is Caregiving Hazardous to One's Physical Health? A Meta-Analysis." *Psychological Bulletin*, 129 (6), 2003, pp. 946–972.

Wagstaff, Adam, and Eddy van Doorslaer. "Income Inequality and Health: What Does the Literature Tell Us?" *Annual Review of Public Health*, 21, 2000, pp. 543–567.

CHAPTER 9

International Justice in Elder Care: The Long Run

Abstract In the short run, the cross-border migration of elder-care workers is a zero-sum game, with the source country losing and the host country gaining. This offends our sense of justice, especially since the host populations tend to be richer. In this chapter, I argue that we ought to direct our gaze beyond the short run, at the long run. Once we do that, we will see possibilities of non-zero-sum games that are mutually beneficial. The large question arises, though, as to how nations may choose among them by committing to some principle of justice.

Keywords Elder care · Migration · Immigration · International justice

1 Introduction

Caring for the elderly is one of the greatest issues of our time. Just a century ago, caring for the elderly was typically a traumatic and brief experience, as the average elderly person would be dying of an acute disease (Lynn 2004). Today, we are much less likely to die of acute diseases. In our later age, we are much more likely to have chronic health problems and long-term disabilities. We are much more likely to need elder care.

This chapter was published as Lee (2011).

L. W. Lee, *Behavioral Economics and Bioethics*,
Palgrave Advances in Behavioral Economics,
https://doi.org/10.1007/978-3-319-89779-0_9

The demand for elder care is widely predicted to grow all over the world. For example, the number of Americans who are 85 and older will double to 9.6 million by year 2030 (DHHS 2003; President's Council 2005, p. 7; Institute of Medicine Committee 2008). At 85 and beyond, 95% of people cannot move around by themselves (Lynn 2004, p. 13). Therefore, without breakthroughs in biomedical technology, more than 9 million Americans will need assistance with basic tasks of living. Their number is projected to double again 20 years later (President's Council 2005, p. 8).

The supply of elder care, on the other hand, does not seem to be growing at a commensurate rate. Spouses—the primary caregivers of the past—are available only in fixed proportion: there are not more of them *per* elderly person, even though there are more *of* them. Also, spouses of elderly people are themselves mostly elderly, with their own needs for assistance. Finally, an elderly person today has fewer children and relatives who may care for them. They are likely to be older, to live farther away, and to have jobs that limit the amount of time they can spend on giving care.

Shortages in elder care in developed countries seem to have been averted largely by immigration (Redfoot and Houser 2005; Browne and Braun 2008). In the United States, estimates of immigrants providing elder care to natives vary with geography and the category of elder care. At one extreme, 95% of "home-care aides" in Hawaii came from the Philippines (Browne et al. 2007). For the United States as a whole, the estimates are around 20% (Smith and Baughman 2007, p. 21). The actual proportion is certainly greater, since some immigrants work informally in private homes.

The dependency on immigrants raises questions of international justice (Eckenwiler 2009). At any moment, the transnational migration of elder-care workers is a zero-sum game: when more young people migrate, inevitably there are fewer of them left in the source population. In terms of the caregiving capacity of a population, the emigration of young people clearly represents a drain. The Philippines, the largest source country for foreign nurses in the United States, is reported to be experiencing shortage of nurses (Lorenzo et al. 2007).

In the long run, however, transnational migration is not a zero-sum game. Migration can cause world populations to change in complicated ways. Over several decades, how a change in migration affects everyone is a difficult question. In this chapter, I show that the migration of young workers may make the world younger, in the sense of all populations having proportionally more young people. In that case, the caregiving burden on the average young person declines all over the world. More migration, therefore, may benefit everyone. This possibility suggests that, when it comes to making ethical judgments about the migration of care workers, we should distinguish between the short run and the long run.

2 Elder Care in the Long Run

The question with which this research began seems simple enough: If young people in one population migrate to another population, what will happen to elder care in these populations? The answer, as it has turned out, is not simple, even when we study the simplest scenario. We must take into account a number of demographic forces.

Imagine a world with two populations, linked by a steady flow of people migrating from one (the source) to the other (the host). In each population, people are either young or elderly. An elderly person requires assistance with basic tasks of living. A young person bears the fair share of the total burden of caregiving in the population, by giving care to elderly persons directly or by paying taxes that are used to buy service in the elder-care market.

A common measure of the burden of elder care in a population is the "dependency ratio": *the average number of elderly persons being supported by a young person.*[1] Consider for a moment the meaning of the dependency ratio. If the dependency ratio is 0.25, then each young person will care for 0.25 of an elderly person, or four young persons will share the cost of caring for an elderly person. If the dependency ratio becomes 0.5, then each young person will care for one half of an elderly person, or two young persons will share the cost of caring for one elderly person. If the ratio is 1.0, then each young person will bear the entire cost of taking care of an elderly person.

The dependency ratio of a population changes when the composition of the population changes. Take the number of elderly people first. There will be more elderly people next year if there are more young people this year: some young people today will become old sooner than others. The rate of "aging" is determined by the age distribution among the young. For example, if the age distribution of the young is uniform

[1]The United Nations (2006, Table A.III.3) defines an elderly person as someone who is 65 years or older and a young person as someone who is between 15 and 64 years old.

between 15 and 64, then the rate of transition from being young to being elderly is equal to 1/50. In addition, the more elderly people there are today, other things being equal, the more of them will die tomorrow.

Now consider what may cause the number of young people in a population to change. Birth, diseases, accidents, and aging come to mind first. The sum of the effects of these factors is "the net replacement rate." Let's say that out of a thousand young people, 20 are born, 5 die of accidents and diseases, and 17 become elderly every year. Then the annual net rate of change equals negative 2, out of a thousand. Second, emigration and immigration obviously matter as well. Third, some young people become ill and die prematurely every year from the burden of caregiving. There is strong evidence that caregiving affects adversely a caregiver's mental and physical health (Kim and Knight 2008; Christakis et al. 2003; Ho et al. 2009). Caregivers are more likely to suffer from depression (Burton et al. 2003; Covensky et al. 2003; Haley et al. 2003; McCusker et al. 2007), stress (Hubbell and Hubbell 2002; Vitaliano et al. 2003; Gaugler et al. 2007) and heart disease (Lee et al. 2003). Caregivers themselves are more likely to need elder care sooner and to die sooner (Schultz and Beach 1999; Christakis and Iwashyna 2003).[2] Thus, there are good reasons for assuming that caregiving has a mortality effect on young people.

The population described above can evolve along any of a large number of time paths, depending on its demographic factors and its interactions with other populations. There are three kinds of time paths: (1) continuous growth, (2) continuous decline (hence eventual collapse), and (3) a stable equilibrium.

I have studied a numerical model of a world with two populations, linked by migration, that reach stable sizes in the long run. (See the Appendix for a mathematical description of the model.) In this world much depends critically on migration. If young immigrants became scarce, the host population will eventually collapse. If elderly immigrants became predominant, the host population will collapse, too.[3] Demographers know that immigration has large effects on the host population. Assuming an annual rate of immigration between 1.4 and 2.1 million (including illegal), Passel and Cohn (2008) project that 82% of

[2] It is also suggestive of the harshness of the work that the turnover rate among care workers in nursing homes is relatively high (Smith and Baughman 2007, pp. 24–25).

[3] There is also the risk that, as a result of population loss, the source population may not be able to sustain itself. The risk is high if the two populations are similar.

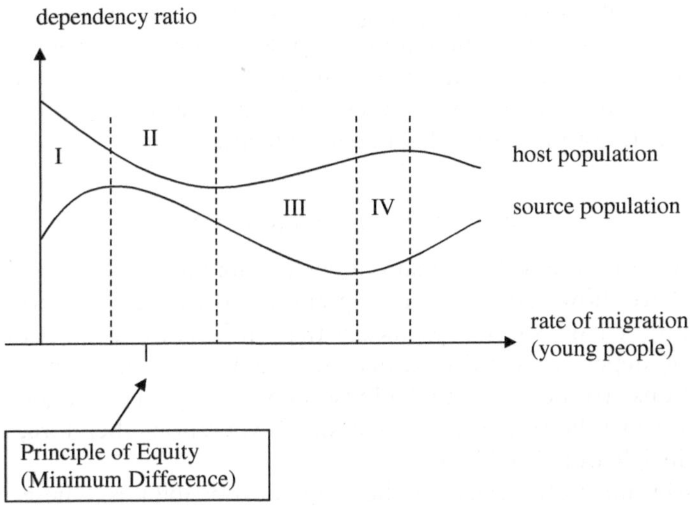

Fig. 1 The long-run effect of migration on elder care

the growth of the U.S. population, from 296 million today to 438 million in 2050, will result from immigrants and their descendants.

The most interesting implication of the model has to do with the long-run behaviors of the dependency ratios. These ratios are functions of the rate of immigration. In the host population, the dependency ratio equals 0.45 when the annual rate of immigration of young people is 0.8 million, and declines to 0.39 when the rate of immigration of young people rises to 0.9 million.[4] So, as expected, the immigration of young people reduces the caregiving burden in the host population. In the source population, the dependency ratio equals 1.0 when it loses 0.8 million young people to emigration every year. Yet, remarkably, the dependency ratio declines to 0.75—meaning proportionally more young people in the population—when it loses young people at the greater rate of 0.9 million. The 25% decrease in the dependency ratio is also more substantial than expected from the 12% increase in migration.

What I have learned from the model can be depicted in Fig. 1. The figure depicts the dependency ratios of the two populations as functions

[4]This relationship does not hold in general. The dependency ratio rises back up to 0.44 when the rate of immigration of young people increases to 1.0 million.

of the rate of migration of young people. The two curves are drawn hypothetically so as to highlight the relative long-run values of the dependency ratios at various rates of migration. Four regions are identified. In region I, the host ratio goes down and the source ratio goes up. In region II, both ratios go down, and so forth.

3 Policy Implications

The long run always holds many more possibilities than the short run. What I have shown above is that migration in the long run is not a zero-sum game. I have shown specifically that an increase in migration may cause the dependency ratios to decline in both the source and the host populations. To the extent that a lower dependency ratio is a good thing, migration may benefit both populations.[5] Migration, therefore, satisfies "the principle of mutual benefits."

This by no means exhausts the long-run possibilities. There may be many migration policies that satisfy the principle of mutual benefits, offering the nations a shot at "the global optimum." The need to choose can be seen clearly in Fig. 1. Phase II contains a whole range of rates of migration that would satisfy the principle of mutual benefits. If the global community can choose the optimum rate of migration within that range, how and what would it choose?

Here I shall note two scenarios. First, the global community can jointly choose what is fair from the set of efficient rates of migration. That means that the joint decision satisfies some notion of equity. For example, perhaps the global community wants to minimize the difference between dependency ratios among members (Anand et al. 2004). As Fig. 1 shows, there always exists a rate of migration that, in the long run, will result in minimum difference between dependency ratios. Of course, the mere possibility of such a global policy does not mean that it would be easy to implement; it would require good data, consensus on demographic forecasts, and, most importantly, the global community's commitment to the principle of equity.

[5]Immigrant caregivers substitute for "informal caregivers," who are spouses and children. Studies show that, as more elder care is bought in the market, informal caregivers' health improves (Christakis and Iwashyna 2003; Gaugler et al. 2007). This is another reason why immigration is good for national health.

Table 1 The two populations in nash solution

Annual rate of migration (millions)	*Dependency ratio*	
	Source population	*Host population*
0.0	0.5	0.7
0.5	0.4	0.6
1.0	0.3	0.5

In the second scenario, members of the global community cannot reach consensus on the basis of a shared principle of justice. Then they will have to negotiate the rate of migration. There are many more possible outcomes here. A well-known outcome is the Nash Solution, which satisfies a number of fairness criteria, including mutual benefits (Luce and Raiffa 1957). Let's consider an example involving two populations that are linked by the migration of young workers as described in Table 1.

Suppose there are only three possible rates of migration: 0.0 million a year, 0.05 million a year, and 1.0 million a year. Should the two populations fail to reach an agreement, there would be no migration and the dependency ratios in both populations would be 0.5 for the source population and 0.7 for the host population. If the annual rate of migration equals 0.5 million, then the long-run dependency ratios will be 0.4 and 0.6, respectively. If the rate equals 1.0 million, then the ratios will be 0.3 and 0.5, respectively. Which of the two rates of migration will be agreed to? The Nash Solution is 1.0 million young workers a year. It is the most efficient because it lowers the dependency ratio of each population most, and it is also fair because it lowers the ratios of both the source and the host populations equally, by 0.2.[6]

[6]The Nash Solution is the solution where the product of the net gains of participants is the greatest. (The largest rectangle that can be formed from a line segment is a square, which has equal borders.) In our case, the gains are the decreases in dependency ratios. With an annual flow of 1.0 million immigrants, the product $(0.3 - 0.5)(0.5 - 0.7) = 0.04$; with an annual flow of 0.5 million immigrants, the product $(0.4 - 0.5)(0.6 - 0.7) = 0.01$. Since 0.04 is greater than 0.01, the annual flow of 1.0 million immigrants is the Nash Solution.

4 Summary

In the short run, the migration of elder-care workers is a zero-sum game. This naturally offends our sense of fairness, especially when the host populations are richer. In this chapter I have argued that we ought to look beyond the short run. Once we do that, we will see possibilities of non-zero-sum games that are mutually beneficial. The large question arises, however, as to how nations may choose among them by committing to some principle of justice.

Appendix: Two Populations Linked by Migration

I assume that there are two populations in the world: the host and the source. In *each* population, the numbers of the elderly (E) and the young (Y) change according to linear differential equations:

$$\frac{\delta E}{\delta t} = aY - dE + i_E \tag{1}$$

$$\frac{\delta Y}{\delta t} = bY - cE + i_Y \tag{2}$$

where

- a the rate of aging
- b the net replacement rate among the young
- c the cost of caregiving in terms of young lives lost per elderly person
- d the death rate among the elderly
- i_E the flow of migration of elderly people (+ if host; – if source)
- i_Y the flow of migration of young people (+ if host; – if source)

In my analysis, I assign numerical values to demographic parameters that ensure that both populations will converge in the long run. See Table 2.

Table 2 Demographic parameters

	a	b	c	d	i_E	i_Y
Host population	0.02	–0.006	0.01	0.05	+200,000	+various
Source population	0.03	0.15	0.11	0.02	–200,000	–various

References

Anand, Sudhir, Fabienne Peter, and Amartya Sen, eds. *Public Health, Ethics, and Equity*. Oxford: Oxford University Press, 2004.

Browne, Colette V., and Kathryn L. Braun. "Immigration and the Direct Long-Term Care Workforce: Implications for Education and Policy." *Gerontology and Geriatrics Education*, 29 (2), July 2008, pp. 172–188.

Browne, C., K. Braun, and P. Arnsberger. "Filipinas as Residential Long-Term Care Providers: Influence of Cultural Values, Structural Inequality, and Immigration Status on Career Choice." *Journal of Gerontological Social Work*, 48, 2007, pp. 698–704.

Burton, Lynda C., Bozena Zdaniuk, Richard Schulz, Sharon Jackson, and Calvin Hirsch. "Transitions in Spousal Caregiving." *The Gerontologist*, 43, 2003, pp. 230–240.

Christakis, Nicholas A., and Theodore Iwashyna. "The Health Impact of Health Care on Families: A Matched Cohort Study of Hospice Use by Decedents and Mortality Outcomes in Surviving, Widowed Spouses." *Social Science and Medicine*, 57 (3), August 2003, pp. 465–475.

Covinsky, Kenneth E., Robert Newcomer, Patrick Fox, Joan Wood, Laura Sands, Kyle Dane, and Kristine Yaffe. "Patient and Caregiver Characteristics Associated with Depression in Caregivers of Patients with Dementia." *Journal of General Internal Medicine*, 18 (12), December 2003, pp. 1006–1014.

Department of Health and Human Services (US). *The Future Supply of Long-Term Care Workers in Relation to the Aging Baby Boom Generation: A Report to Congress*. Washington, DC, 2003.

Eckenwiler, Lisa. "Care Worker Migration and Transnational Justice." *Public Health Ethics*, 2 (2), July 2009, pp. 171–183.

Gaugler, Joseph E., Anne M. Pot, and Steven H. Zarit. "Long-Term Adaptation to Institutionalization in Dementia Caregivers." *The Gerontologist*, 47 (6), 2007, pp. 730–740.

Haley, William E., Laurie A. LaMonde, Beth Han, Allison M. Burton, and Ronald Schonwetter. "Predictors of Depression and Life Satisfaction Among Spousal Caregivers in Hospice: Application of a Stress Process Model." *Journal of Palliative Medicine*, 6 (2), April 2003, pp. 215–224.

Ho, Suzanne, Alfred Chan, Jean Woo, Portia Chong, and Aprille Sham. "Impact of Caregiving on Health and Quality of Life: A Comparative Population-Based Study of Caregivers for Elderly Persons and Noncaregivers." *The Journals of Gerontological Series A: Biological Sciences and Medical Sciences*, 64A (8), 2009, pp. 873–879.

Hubbell, Larry, and Kelly Hubbell. "The Burnout Risk for Male Caregivers in Providing Care to Spouses Affiliated with Alzheimer's Disease." *Journal of Health Human Service Administration*, 25 (1), Summer 2002, pp. 115–132.

Institute of Medicine Committee on the Future Health Care Workforce for Older Americans. *Retooling for an Aging America: Building the Health Care Workforce*. Washington, DC: National Academies Press, 2008.

Kim, Jung-Hyun, and Bob G. Knight. "Effects of Caregiver Status, Coping Styles, and Social Support on the Physical Health of Korean American Caregivers." *The Gerontologist*, 48, 2008, pp. 287–299.

Lee, Li Way. "International Justice in Elder Care: The Long Run." *Public Health Ethics*, 4 (3), 2011, pp. 292–296.

Lee, Sunmin, Graham A. Colditz, Lisa F. Berkman, and Ichiro Kawachi. "Caregiving and Risk of Coronary Heart Disease in U.S. Women: A Prospective Study." *American Journal of Preventive Medicine*, 24 (2), February 2003, pp. 113–119.

Lorenzo, F. M. E., J. Galvez-Tan, K. Icamina, and L. Javier. "Nurse Migration from a Source Country Perspective: Philippine Country Case Study." *Health Services Research*, 42, June 2007, pp. 1406–1418.

Luce, R. Duncan, and Howard Raiffa. *Games and Decisions*. New York: Wiley, 1957.

Lynn, Joanne. *Sick to Death and Not Going to Take It Anymore!* Berkeley: University of California Press, 2004.

McCusker, Jane, Eric Latimer, Martin Cole, Antonio Ciampi, and Maida Sewitch. "Major Depression Among Medically Ill Elders Contributes to Sustained Poor Mental Health in Their Informal Caregivers." *Age and Ageing*, 36 (4), 2007, pp. 400–406.

Passel, Jeffrey, and D'Vera Cohn. "U.S. Population Projections: 2005–2050." PEW Research Center, February 2008.

Redfoot, Donald L., and Ari N. Houser. *We Shall Travel On: Quality of Care, Economic Development, and the International Migration of Long-Term Care Workers*. Washington, DC: AARP Public Policy Institute, October 2005.

Schultz, Richard, and Scott Beach. "Caregiving as a Risk Factor for Mortality." *Journal of the American Medical Association*, 282, 1999, pp. 2215–2219.

Smith, Kristin, and Reagan Baughman. "Caring for America's Aging Population: A Profile of the Direct-Care Workforce." *Monthly Labor Review*, September 2007, pp. 20–26.

The President's Council on Bioethics. *Taking Care: Ethical Caregiving in Our Aging Society*. Washington DC, September 2005.

United Nations. *World Population Ageing 2007*. Population Division, 2006.

Vitaliano, Peter P., Jianping Zhang, and James M. Scanlan. "Is Caregiving Hazardous to One's Physical Health? A Meta-Analysis." *Psychological Bulletin*, 129 (6), 2003, pp. 946–972.

PART IV

People and Animals

CHAPTER 10

The Making of Modern Cruelty

Abstract Modern cruelty to animals is both more extensive and more intensive than it was before Industrial Revolution. I attribute these trends to the ascendancy of distancing institutions (such as slaughterhouses and meat-packing plants) and the growing capacity for willful blindness. As we continue to specialize in tasks, find more distancing institutions, and invent ways of promoting willful blindness, we grow more oblivious to our cruelty to animals. We continue to weaken the link from cruelty to compassion, thereby inflicting more cruelty on animals. Further, when there is more cruelty, there is more incentive to promote willful blindness. Cruelty and blindness feed on each other.

Keywords Distancing · Willful blindness · Cruelty

1 The Suppression of Compassion

I am visiting animals now. For a long time, I have noticed something strange in our relation with animals: We treat animals with cruelty, even as we have compassion for them. How do we manage this contradiction?

The answer, I think, lies in the very nature of compassion. As Adam Smith (1759, Chapter 1) observes, compassion is derived from the suffering of something that we can sympathize with. It is an uncomfortable, negative emotion. To avoid feeling compassion, we can reduce either the

L. W. Lee, *Behavioral Economics and Bioethics*,
Palgrave Advances in Behavioral Economics,
https://doi.org/10.1007/978-3-319-89779-0_10

suffering or our perception of the suffering. We find it much easier to reduce our perception of the suffering than to reduce the suffering.

This explains two proverbs that my school teacher told us to memorize (because they would be on the exam):

1. An ethical man stays away from the kitchen: brutality is perpetrated there every day.
2. He who hears its scream will lose his appetite for its flesh.

The two proverbs, more than any others, still ring in my ears today. They strike me now as simply admonitions to conserve compassion.

As I visit animals in this chapter, I pay particular attention to the ways that we have found to suppress compassion. They are grouped into two categories: "distancing institutions" and "willful blindness."

2 Distancing Institutions

We have invented many "distancing institutions" (Bandura 1999). These are devices that keep us away from animals, in order to dull our compassion for them. We increase our distance from them by devising new products, new processes, and new institutions. An example is the kitchen of the house I grew up in. A very heavy door separated it from the rest of the house. The kitchen was two steps down from the rest of the house. It was dark. And it had its own door out to the street. Kids and men just knew to stay away from the kitchen.

Another example is the modern animal farm. We build animal farms far away from us and we build them with few windows, so we do not see the animals in them. Other familiar "distancing institutions" are slaughterhouses and meat processors. The MacDonald's down the road is a distancing institution, too: It keeps animals out of our sight and mind until we are seated and have ketchup and napkins ready.

The Industrial Revolution greatly accelerated distancing. Specialization allows consumers to avoid unpleasant tasks, which are taken over by producers. Overtime, consumers become blind to them. Also, the rise of Madison Avenue allows producers to use mass media to create images that divert consumers' attention (Lee 2015).

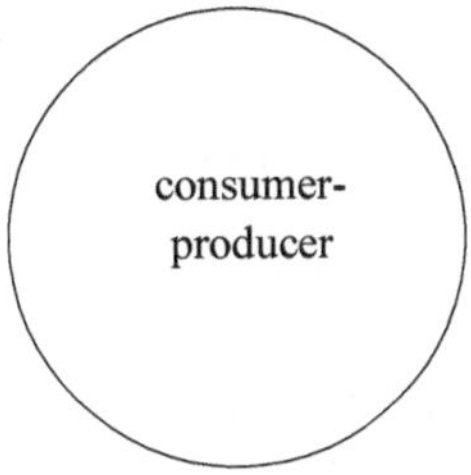

Fig. 1 Robinson Crusoe

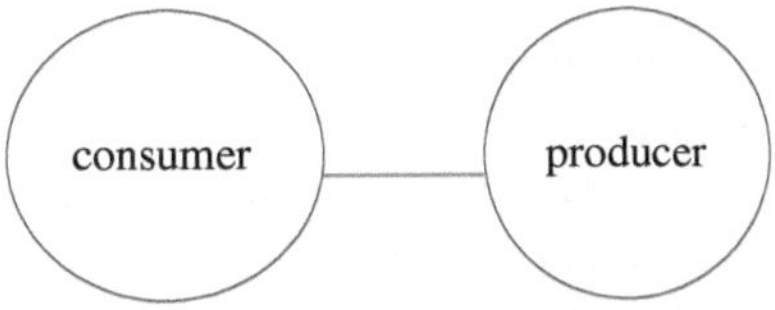

Fig. 2 One degree of separation in apples

Let's illustrate distancing. In the beginning, consumption and production were integrated. We were all Robinson Crusoes. Figure 1 is a picture of Robinson Crusoe.

Then, we found markets, where we could buy things that we did not produce. Somehow, we overcame "the endowment effect" (Thaler 1980; Kahneman 2011); we began to consume things we did not produce ourselves. Take apples, for example. I eat apples, even though I don't have an apple tree in my backyard. So my consumption is separated from the production. In Fig. 2, the degree of separation is equal to one.

Consider a pig's journey from the factory farm to our dinner plate. In Fig. 3, the degree of separation of a consumer from the pig farmer is equal to three and from the slaughterhouse equal to two. A consumer of pork knows less about how a pig lives in a stall on an animal farm than a consumer of apple knows about how an apple grows on a tree.

There are several excellent accounts of distancing. Maccoby (1983) describes how executioners (of people) adopt a detached attitude. Serpell (1996) gives a historical account of the success we have had in developing distance between ourselves and animals. Bandura (1999) and Heffernan (2011) point their fingers at the division of labor (i.e., specialization) as a contributor to distancing.

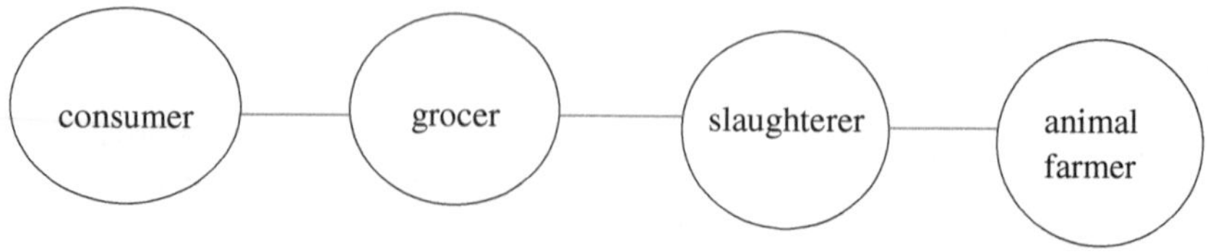

Fig. 3 Three degrees of separation in pork

3 Willful Blindness

Along with distancing institutions, we make plenty use of willful blindness. This blindness is our choice to *not* see certain things—or in the words of Bazerman and Tenbrunsel (2011)—our choice to do "psychological cleansing" by "moral disengagement." Frank (2006) tells us about the "ignorance-is-bliss" phenomenon at animal slaughterhouses. Joy (2010) explains people's ability to develop "psychic numbing" to cope with the cognitive dissonance in petting dogs while eating pork chops. Pachirat (2011) tells us his experience at slaughterhouses, where even workers at the killing floors become oblivious to the gruesome cruelty being inflicted on cows.[1]

While no doubt willful blindness increases with the degree of separation of consumption and production, I believe that another driver of willful blindness is complexity. There is strong evidence of willful blindness in financial markets, where products are extremely complex and they create bubbles and panics. An example is the mortgage derivative that is composed of thousands of subprime mortgages. And some people have profited from the realization that, the more complex is the derivative, the more buyers there are and the more they are willing to pay for it.

Let's illustrate how complexity promotes blindness to cruelty. Imagine that you are considering whether to make a birthday cake. It has dozens of ingredients, including milk and eggs. It is a more complex product than the proverbial grandma's chicken soup. Figure 4 captures the complexity of a cake:

[1]Willful blindness is evident in our languages, too. Balcombe (2016) prefers the word "fishes" to the plural "fish." Fishes are animals, too. By the same token, I am trying to remember to write and say the plural "sheep" as "sheeps," though spelling check always corrects me instantly.

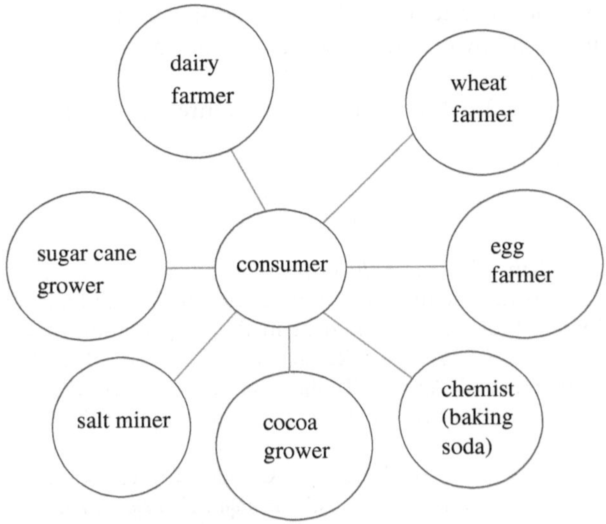

Fig. 4 Seven degrees of complexity in a birthday cake

Since there are seven producers involved in the cake, you do not pay all your attention to the dairy farmer or the egg farmer or any of the other producers. So your attention to any particular ingredient is very thin. You may very well employ willful blindness as a heuristic to decide whether to make the cake or not.

4 The Future of Cruelty

As we continue to specialize in tasks, find more distancing institutions (such as slaughterhouses and meat packers) and invent ways of promoting that veil of ignorance (such as the Madison Avenue), we grow more oblivious to our cruelty to animals. We continue to weaken the link from cruelty to compassion, thereby inflicting more cruelty on animals. Further, when there is more cruelty, there is more incentive to promote willful blindness. Cruelty and blindness tend to feed on each other.

Our ability to manipulate blindness of ourselves and others makes for a short half-life for any justice achieved. Distancing has become part and parcel of the technological change that drives the wedge between consumption and production. At the same time, complexity has taken on a business life of its own in modern economies. Merchants of complexity are doing well, thank you.[2]

In my next visit, I will be looking for countervailing forces against our cruelty to animals.

References

Balcombe, Jonathan. *What a Fish Knows: The Inner Lives of Our Underwater Cousins*. New York: Scientific American/Farrar, Straus, and Giroux, 2016.

Bandura, Albert. "Moral Disengagement in the Perpetuation of Inhumanities." *Personality and Social Psychology Review*, 3 (3), 1999, pp. 193–209.

Bazerman, Max H., and Ann E. Tenbrunsel. *Blind Spots: Why We Fail to Do What's Right and What to Do About It*. Princeton, NJ: Princeton University Press, 2011.

Frank, Joshua. "Process Attributes of Goods, Ethical Considerations and Implications for Animal Products." *Ecological Economics*, 58, 2006, pp. 538–547.

Heffernan, Margaret. *Willful Blindness: Why We Ignore the Obvious at Our Peril*. New York: Walker and Company, 2011.

Joy, Melanie. *Why We Love Dogs, Eat Pigs and Wear Cows: An Introduction to Carnism*. San Francisco: Conari Press, 2010.

Kahneman, Daniel. *Thinking, Fast and Slow*. New York: Farrar, Straus, and Giroux, 2011.

Lee, Li Way. "Ethical Consumption." In Morris Altman, ed. *Real-World Decision Making: An Encyclopedia of Behavioral Economics*. Santa Barbara: ABC-CLIO, LLC, 2015.

Maccoby, Hyam. *The Sacred Executioner: Human Sacrifice and the Legacy of Guilt*. London: Thames and Hudson, 1983.

Pachirat, Timothy. *Every Twelve Seconds: Industrialized Slaughter and the Politics of Sight*. New Haven, CT: Yale University Press, 2011.

Serpell, James. *In the Company of Animals*. Cambridge, Great Britain: Cambridge University Press, 1996.

Smith, Adam. *The Theory of Moral Sentiments*. Indianapolis, IN: Liberty Classics edition, 1759/1976.

Thaler, Richard H. "Toward a Positive Theory of Consumer Choice." *Journal of Economic Behavior and Organization*, 1, 1980, pp. 39–60.

Wu, Tim. *The Attention Merchants*. New York: Alfred A. Knopf, 2016.

[2]Merchants of complexity sell inattention; they are in the same business as the "merchants of attention" (Wu 2016).

CHAPTER 11

Two Animal Ethics; Many More Economic Lessons

Abstract Cruelty to animals is a case of the tragedy of the commons. Economics has been tackling this type of problems for a long time. The experience is mixed: some successful and some disastrous. The successes and the failures hold lessons for animal advocacy. To make the case, I consider the two major animal ethics—"animal rights" and "animal welfare." And I argue that animals need them both.

Keywords Animal rights · Animal welfare · Animal advocacy

1 Introduction

Cruelty to animals is a case of the tragedy of the commons (Hardin 1968; Cowan 2006). Most of us recoil at the thought of it. But when lunch time comes, we eat ham sandwiches without much of a second thought. Then we complain about cruelty again. We do know one thing: To fix the problem, we need to take actions collectively. We must have policies that make each of us pay the cost of cruelty to animals.

Economics has been tackling tragedies of the commons for a long time. The record is mixed. The successes and the failures hold lessons for animal advocacy. To see what the lessons are, I consider the two major

L. W. Lee, *Behavioral Economics and Bioethics*,
Palgrave Advances in Behavioral Economics,
https://doi.org/10.1007/978-3-319-89779-0_11

animal ethics—"animal rights" and "animal welfare." Both are active in animal markets. And I argue that animals need them both.

2 Animal Rights as a Demand-Side Ethic

Animal-rights ethic says that all sentient beings have the right not to suffer. Animal-rights advocates oppose all forms of demand for animals, including medical experimentation and entertainment. Gary Francione (2014) explains: "The abolitionist approach sees the problem of animal exploitation primarily as one of demand and not supply. That is, the problem is not that there are institutional exploiters who will provide animal products to the public; the primary problem is that the public demands those products."[1]

From that vantage point, animal-rights advocacy is similar to public policies toward alcohol, tobacco, gambling, and firearms. All these policies are aimed directly at reducing ***demand***, not supply. Behavioral economists have learned, for example, that consumers who are being pressed to change their choices are prone to employ evasion tactics. We can see these tactics in the ways by which consumers react to the push for humane diets. Vegetarianism has splintered into lacto-ovo vegetarianism, lacto vegetarianism, ovo vegetarianism, etc. (Wikipedia, "Vegetarianism"). Even "semi-vegetarianism" has several branches, including flexitarianism, pescetarianism, and pollotarianism (ibid.). Another evasive tactic is substitution. Consumers who are forced to go through meatless Mondays may gorge on meat on Sundays and Tuesdays.

Behavioral economists also have learned that consumers do not always respond to more and better information about a product. Consumers respond more predictably to "textured information." Richard Thaler and others have shown how to make people save more (Sunstein and Thaler 2003; Thaler and Benartzi 2004). They designed "saving nudges," which are default mechanisms. A company may automatically withhold 5% of an employee's paycheck and deposit it in a retirement account—unless the employee expressly chooses to opt out. (This is different from the prevailing system, where deduction is made for retirement only if an employee expressly chooses to opt in.)

[1] The primary animal-rights group in the US today is Mercy for Animals.

When it comes to reducing the demand for meat, a nudge would be a cafeteria plan that offers vegan spaghetti sauce as the default, so that if you want to have spaghetti sauce with meat, you must ask for it. Most people probably won't bother to ask.

3 Animal Welfare as a Supply-Side Ethic

Animal-welfare advocates want to minimize suffering in the aggregate. When they initiate ballot initiatives against veal crates and gestation stalls, they intend to prohibit certain methods of raising animals. When they expose workers abusing animals in factory farms and slaughterhouses, they intend to galvanize actions to pressure producers into abolishing cruel practices. They strive to change "the animal production function," which is on the supply side of the market. Animal-welfare ethic is a supply-side ethic. It sees suffering in how an animal is raised and how many are raised.[2]

Economics suggests two supply-side strategies: (1) raise cost and (2) monitor compliance.

3.1 Raise Cost

Factory farms do not pay the full social cost of the cruelty in their methods, just as power plants do not pay the full social cost of burning fossil fuels. The analogy to pollution suggests ways of reducing cruelty. One way is what may be called "cap-and-trade cruelty permits." The idea here is that, if a farm does not meet a certain standard of humaneness, the farmer must pay for it by purchasing permits on the open market. This market works much like the market for "pollution permits," where polluters buy and sell permits. Advances in animal science have made it possible to calibrate the extent of suffering on each animal, so a producer will know how many permits to buy or sell.[3] The producer who treats

[2]The primary animal-welfare group in the US today is The Humane Society of the United States.

[3]A real-world example of a calibration system for cruelty is that administered by Global Animal Partnership. The system has five "steps," where a higher number means a less cruel method: Step 1: no crates, no cages, no crowding; Step 2: enhanced indoor environment; Step 3: outdoor access; Step 4: pasture centered; Step 5: animal centered; no physical alternations; Step 5+: entire life on same farm with on-site or local slaughter.

animals well can sell the permits that he does not need; the producer who mistreats animals must buy the permits that he is required to have but does not have.

Another cost issue is the subsidization of corn and soybean. Because they are subsidized, the price of animal feed that contains them is lower than it would be otherwise. Low feed price leads to more animals being raised. Without the subsidies, the cost of animal products will be higher, and so the supply will be lower. The same reasoning applies to local tax breaks for factory farms.

I see another way to reduce cruelty on the supply side that does not involve taxes or permits: Enhance producers' market power. Exempt animal producers from all antitrust policies. Encourage them to merge for monopolization. Higher prices will serve animals well.

3.2 Monitor Compliance

The supply side of a market is harder to control than the demand side. This is a lesson learned from attempts to control the production and distribution of alcohol in the 1930s and cocaine in the 1980s. Most economists would say that those attempts failed miserably. Cruelty is more difficult to detect than alcohol or cocaine, and that is an additional hindrance to the enforcement of policies against cruelty.

Last but not least, we must plug the loophole by the name of free trade. When we examine free-trade meat, we cannot easily tell the degree of cruelty in the methods of foreign factory farms. What is more, prevailing trade agreements are rife with exemptions, with the result that foreign producers have no incentive to recognize the moral cost of cruelty (Matheny and Leahy 2007).[4]

4 Do We Need Both Ethics?

The answer is yes, and the reason has to do with another economic lesson: A change in demand can cause a change in supply, and vice versa. This is most easily explained by an example.

[4]Free trade includes commerce within a country. California has effectively blocked free trade by requiring that eggs sold in that state must come from hens that are raised in larger cages than those in most other states.

Fig. 1 Fewer hens in smaller cages

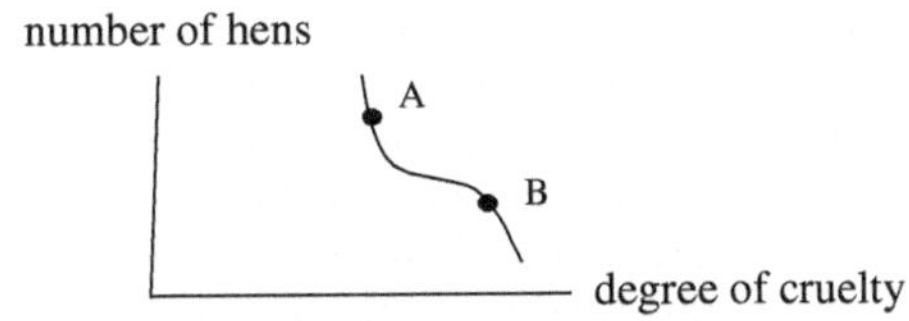

Suppose that a group of consumers become vegan, in support of animal rights. Immediately, the demand for eggs goes down, and fewer eggs will be produced. That is good from the point of view of animal rights, since fewer laying hens will need to be raised to suffer. Still, some hens will continue to be raised to meet the demand of other consumers, and the question arises as to how these hens will fare. Will they live in larger or smaller cages than before? Will they be "free-range"? If so, will they be protected from eagles and foxes? Our intuitive understanding of "mass production" seems to make us think that fewer hens would go hand in hand with more humane production methods. But we cannot depend entirely on our intuition here: there is high stake in the specifics of demand and supply. When demand drops, individual owners of factory farms have to lower price to sell the same number of eggs and they make less profit. They may cut back on the production of eggs by raising fewer hens. This does not mean that the hens, now fewer, would have more room to spread wings in those same old cages. The producers may use even smaller cages to cut cost. In that case, the vegan movement reduces the number of hens but raises the degree of cruelty. Figure 1 illustrates this tradeoff. As the market moves from A to B, fewer hens suffer but the hens that continue to be raised will suffer more than before.

As Fig. 1 may be surprising, I have made up a story about it.

A Story about Fig. 1

There are two consumers in the market for chicken (meat). Each is willing to pay $5 for a chicken raised in a battery cage and $10 for one raised on pasture. Also, each consumer eats one chicken a week regardless of price.

The two chickens are raised by a single producer. To use the battery-cage method, there is the cost of a cage at $2, and the cost of feed at $2 per chicken. To use the pasture method, the cost of the pasture is $6 and the cost of feed is $4 per chicken. Both the cage and the pasture are big enough for two chickens.

Now, if we do a little bit of math, we can figure out what it will cost to raise two chickens: \$6 when using a battery cage and \$14 when using pasture. So the profit, which is the difference between what the two chickens can sell for and the cost of raising them, is \$4 (= \$5 × 2 – \$6) from using the battery-cage method and \$6 (= \$10 × 2 – \$14) from the pasture method. Clearly, it is more profitable to use the pasture system for two chickens. So the market with two chickens is pretty humane.

Suppose that, newly convinced of animal rights, one of the two consumers stops eating meat. Then only one chicken is demanded in the economy. Now, which method is the more profitable to use to raise one chicken? A little bit of math shows that the profit from using the battery-cage method is \$1 (= \$5 – \$4), while that from using the pasture method is zero (= \$10 – \$10). So, for an economy with one chicken, it is more profitable to use the battery-cage method. That is a relatively inhumane way of raising a chicken. And that is a step backward ethically compared with the method earlier when more consumers ate more chickens.

***

This story is about an anomaly when only one of the two ethics is active. The moral of the story, therefore, is that the anomaly is less likely when both ethics are active. Assume that, at the same time when animal-rights groups promote veganism, animal-welfare groups pressure chicken factory farmers to use larger cages. Then animal rights and animal welfare would reinforce each other, instead of working at cross purposes. Figure 2 depicts the market moving from A to B: fewer hens and less cruelty. Incidentally, the producer would break even.

Yes, we need both ethics. The animal-rights ethic reduces the demand for animals, while the animal-welfare ethic directs factory farms toward more humane methods.

Fig. 2 Fewer hens in larger cages

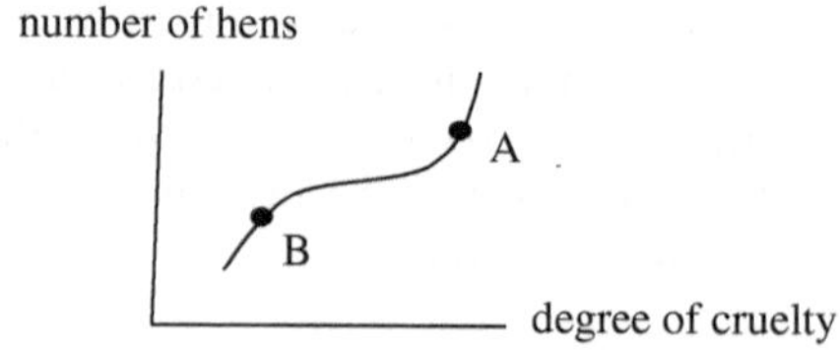

5 A Concluding Remark

At the end of this visit with animals, I have mixed feelings. I see our relationship with them as incredibly extensive and deep; I see our relationship with them as incredibly unjust; I see a ray of hope in the incredibly selfless advocacies for animals, whether for animals' rights or for animals' welfare.

References

Cowen, Tyler. "Market Failure for the Treatment of Animals." *Society*, 43 (2), 2006, pp. 39–44.

Francione, Gary. "Abolitionist Animal Rights/Abolitionist Veganism: In a Nutshell." 2014. Accessed at http://www.abolitionistapproach.com/abolitionist-animal-rights-abolitionist-veganism-in-a-nutshell/#more-9946.

Hardin, Garrett. "The Tragedy of the Commons." *Science*, 162, 1968, pp. 1243–1248.

Matheny, Gaverick, and Cheryl Leahy. "Farm-Animal Welfare, Legislation, and Trade." *Law and Contemporary Problems*, 70, 2007, pp. 325–358.

Sunstein, Cass, and Richard Thaler. "Libertarian Paternalism." *American Economic Review*, 93, 2003, pp. 175–179.

Thaler, Richard, and Shlomo Benartzi. "Save More Tomorrow™: Using Behavioral Economics to Increase Employee Saving." *Journal of Political Economy*, 112 (S1), 2004, pp. 164–187.

CHAPTER 12

Revenges of the CAFO Pigs

Abstract People raise pigs in concentrated-animals-feeding operations (CAFOs). On surface, people's dominance appears to be absolute and complete. But the dominance comes at the cost of declines in people's health and longevity and fertility. I show that there is justice coming out of CAFOs after all.

Keywords Interspecies justice · Factory farming · CAFO

1 Introduction

For what they do for people, pigs have gotten a rotten deal. They live and die in factories known as concentrated-animal-feeding operations, or CAFOs (Imhoff 2010). These are hellish places. As Serpell (1996, pp. 9–11) puts it, lucky pigs get slaughtered before their first birthday, while the unlucky ones become sows. It is hard to imagine justice at CAFOs.

In this chapter, I visit a community of people and CAFO pigs. On surface, people appear to dominate pigs. Taking a deeper look, I find reasons to believe that the dominance comes at a price. It looks as if pigs are capable of revenges; it looks as if there is no free cruelty to pigs after all.

L. W. Lee, *Behavioral Economics and Bioethics*,
Palgrave Advances in Behavioral Economics,
https://doi.org/10.1007/978-3-319-89779-0_12

During my visit, I pay particular attention to the dynamics among people and pigs.[1] Pigs do not openly revolt against people's cruelty toward them. However, they spell trouble for people's welfare over time. For example, I find a reason why people's attempt to grow their own population will be unsuccessful in the long run. And that is because people eat pigs. Pigs do not intend to revenge; nonetheless, they revenge.

2 The Community of People and Pigs

People and pigs form a community because they depend on each other. Their interdependence may be depicted as a feedback loop as follows:

2.1 The People Population

Figure 1 shows that the people population changes for two reasons: people themselves and pigs.

On their own, people reproduce, grow old, and die. Let's say that these forces alone would result in growth in people's population (see arrow 1). Now people also die because they live with pigs. For one thing, they eat a diet with pork. We know today that pork is not altogether a healthy food. Compared with other sources of protein, it has higher levels of antibiotics, cholesterol, and saturated fat. Pork increases risks of liver failure, heart failure, and stroke (see arrow 2).

The pork content of people's diet at a point in time is measured by the ratio of pigs and people, as in Fig. 2.

(Have you noticed the tail in *Q*, and the raised arms in *Y*?)

2.2 The Pig Population

Now I turn to the pig population. Figure 1 shows that it changes also for two reasons: pigs themselves and people.

In this community, people and pigs are completely segregated. All pigs live the entire life at CAFOs. The conditions there are totally unnatural to them. As a result, pathogens break out at CAFOs, killing them (Akhtar 2012). Without counter measures, the pig population would decline (see arrow 3).

[1]For other views on animals-and-people communities, see Blackorby and Donaldson (1992) and Clarke and Ng (2006).

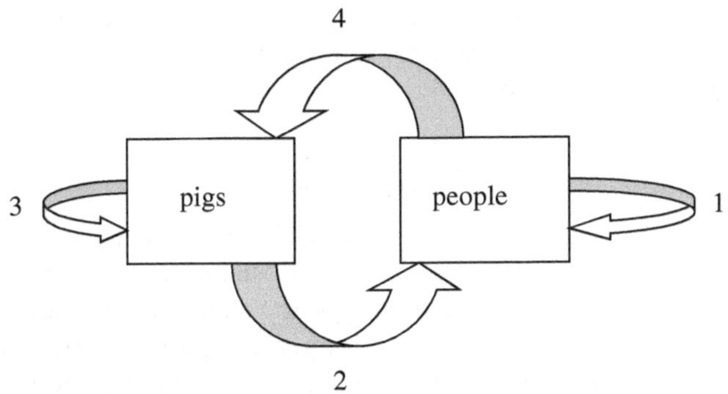

Fig. 1 The community

Fig. 2 The pork content of people's diet

$$\frac{Q}{Y}$$

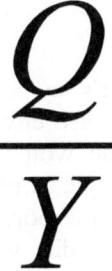

But people have figured out ways to stem the decline. They have come up with antibiotics for controlling the pathogens, and also methods of artificial insemination. I also observe that people have irrepressible desire for pork. The more people there are, the faster people raise pigs. The pig population grows with the people population (see arrow 4).

3 Revenges of the Pigs

When I began my visit, I saw no justice for CAFO pigs at all. Then I started to look deeper below the surface, on the hunch that there may be dynamic justice lurking down there. That is, maybe people who take advantage of pigs now will regret later? That is the question I ask now.

I assume that people consider four things as important to their welfare: their longevity, their health, their palate pleasure, and their pocketbook. Let's consider, respectively, the consequences of four policies designed by people to promote them. Each policy takes the form of a stronger or weaker link in Fig. 1.

3.1 People's Longevity

Let's say people have decided to provide for better elder care at nursing homes. The new policy is intended to boost longevity, and it does not restrict diet in any way. (In Fig. 1, the policy makes for stronger arrow 1.) One would think that, as people live longer, the people population would grow. Yet, the policy will ultimately result in a decline in people's population. The reason has to do with a longer life enabling a greater amount of pork to be consumed.[2] Meanwhile, there are fewer pigs suffering in CAFOs. Have the pigs revenged?

3.2 People's Health

Let's say people start to take a new anti-cholesterol pill that reduces the adverse effects of pork. (In Fig. 1, the pill makes for weaker arrow 2.) One would think that the people population would grow, and that people could eat more pork. But these things will not happen. The people population shall decline; furthermore, the pork content in people's diet will remain constant. Meanwhile, with the pig population in decline as well, there are fewer pigs suffering in CAFOs. Have the pigs revenged?

3.3 People's Pocketbook

Let's suppose that people have found a vaccine that slows down the spread of pathogens deadly to CAFO pigs. (In Fig. 1, the vaccine weakens arrow 3.) The pig population shall grow. But so shall the people population, though at a lower rate. The result is that people end up having higher pork content in their diet. Their health suffers. Have the pigs revenged?

[2]The policy reduces the pig population at the same rate as it reduces the people population. It follows that the pork content of people's daily diet shall remain unchanged.

3.4 People's Palate Pleasure

Let's suppose that people are developing a greater desire for pork, as the result of persuasive advertising for "the other white meat," complete with new recipes. So the demand for pork goes up. (In Fig. 1, the campaign makes for stronger arrow 4.) And more pigs are raised to meet the greater demand. While the pig population grows, the people population grows as well, though more slowly. In the end, people's diet will have higher pork content. People's health declines. Have the pigs revenged?

4 Conclusion

In the end of my visit to this community, I find justice. When people try to multiply, there will be fewer people; when people try to become healthier, they will become sicker; when people try to raise more pigs, there will be fewer pigs. These are signs of dynamic justice at work.

Incidentally, I expect the same sort of dynamic justice to prevail for other animals in CAFOs: meat chickens, egg chickens, meat cows, dairy cows, ducks, turkeys, and fishes.

Appendix

A Formal Model of the Community of Pigs and People

What will happen in this community of pigs and people over time? Will their populations grow steadily? Or will they become extinct? It is impossible to tell from Fig. 1. So let's look at a special case instead. Let the letter Q stand for pigs, and the letter Y for people, and t for time:

Q pigs
Y people
t time

The community of people and pigs is described by two equations of motion:

$$\frac{dY}{dt} = a_1 Y - a_2 \frac{Q}{Y} \tag{1}$$

$$\frac{dQ}{dt} = b_1 Y - b_2 Q \tag{2}$$

The *a*'s and the *b*'s correspond to the arrows in Fig. 1:
arrow 1 = a_1 rate of natural growth of people
arrow 2 = a_2 rate of death caused by diet with pork
arrow 3 = b_1 rate of growth in demand for pork
arrow 4 = b_2 rate of death of pigs due to pathogens

This community will become stable eventually, with populations equal to:

$$Y = \frac{a_2 b_1}{a_1 b_2} \tag{3}$$

$$Q = \frac{a_2 (b_1)^2}{a_1 (b_2)^2} \tag{4}$$

Note that the pig population always moves in lock steps with the people population whenever their environment is perturbed. Divide Q by Y and get:

$$Q/Y = b_1/b_2 \tag{5}$$

This ratio—the number of pigs per person—measures the pork content of people's diet. This ratio is the key to understanding pigs' ability to revenge for the cruelty of CAFOs.

References

Akhtar, Aysha. *Animals and Public Health: Why Treating Animals Better Is Critical to Human Welfare*. New York: Palgrave Macmillan, 2012.

Blackorby, Charles, and David Donaldson. "Pigs and Guinea Pigs: A Note on the Ethics of Animal Exploitation." *Economic Journal*, November 1992, pp. 1345–1369.

Clarke, Matthew, and Yew-Kwang Ng. "Population Dynamics and Animal Welfare: Issues Raised by the Culling of Kangaroos in Puckapunyal." *Social Choice Welfare*, 27, 2006, pp. 407–422.

Imhoff, Daniel, ed. *CAFO: The Tragedy of Industrial Animal Factories*. Foundation for Deep Ecology, Sausalito, CA, in collaboration with Earth Aware, San Rafael, CA, 2010.

Serpell, James. *In the Company of Animals: A Study of Human–Animal Relationships*, 2nd ed. Cambridge: Cambridge University Press, 1996.

PART V

Present People and Future People

CHAPTER 13

Future Earth: A View from the Rainbow Bridge

Abstract We reach out to future generations by mitigating climate risk to them, while they reach us by persuading us to have compassion for them. Both mitigation and persuasion are, therefore, critical to their well-being and our own well-being. They form a feedback loop. As I look closely at the feedback loop, I become concerned that we tend to emphasize mitigation at the expense of persuasion.

Keywords Climate change · Future generations · Mitigation Persuasion

1 Introduction

In this chapter, I visit Future Earth. It is not here now, I know. But we always treat it as if it were. Take the people on Future Earth, for example. We cannot touch them, nor they us. We don't know who they are, or how many they are, or when they will show up in flesh and blood. Nonetheless, they exist in our imagination; we can see them in our own images.

Future Earth looms large at this time, the beginning of an era that we call Anthropocene. As the surface of Earth warms up, we imagine lives becoming more difficult, with more frequent and more severe storms, floods, wildfires, food shortages, and wars. Future generations will be

L. W. Lee, *Behavioral Economics and Bioethics*,
Palgrave Advances in Behavioral Economics,
https://doi.org/10.1007/978-3-319-89779-0_13

suffering in proportion to the risk of these calamities. We feel responsible for the greater risk and sympathetic for the greater suffering.

But it is not just sympathy that we feel; it is compassion. Sakurai (2009) sees distinctions among compassion, pity, sympathy, and empathy. Burton (2015) defines compassion as "suffering with" and observes that it is "associated with an active desire to alleviate the suffering of its object." That is to say, compassion prompts us to take actions to reduce the suffering that gives rise to it. We offer food to a dog in starvation, extend our hands to someone slipping off a cliff, take time to give directions to strangers in town, and give money to people poorer than we are. In all these instances, compassion gives us the will to act to help the less fortunate. Without the will to act, we would do nothing, even if we had the means to do a great deal. Frank (1988) calls sympathy a "commitment device."

In short, our compassion and their suffering are interdependent: our compassion affects their suffering, and their suffering affects our compassion. The two emotions form a feedback loop. This feedback loop determines what justice prevails in the future world. When Dasgupta et al. (1999) describe intergenerational justice as a "recursive relation," and when Dietz and Stern (2008, p. 94, 107) argue that ethics and climate have "simultaneous importance," they probably have that feedback loop in mind. I will call it "the compassion loop."

2 The Compassion Loop

Figure 1 depicts the compassion loop. There are two links. Link "persuasion" suggests that we must be persuaded of climate risk so we develop compassion toward future generations. We feel more compassionate toward future generations when we are persuaded that they face a greater climate risk. Link "mitigation" suggests that our compassion drives our actions to reduce climate risk. The greater is our compassion, the more we will do about the risk.

The two links—"persuasion" and "mitigation"—are both crucial to the workings of the compassion loop. Imagine a loop without either persuasion or action. Then there would be no interaction between compassion and climate risk. Without action, even the strongest persuasion would not make a dent on climate risk. Without persuasion, even the best technology would sit idly by, for there would be no motivation for using it.

Let's take a closer look at the structure of the loop.

Fig. 1 The compassion loop

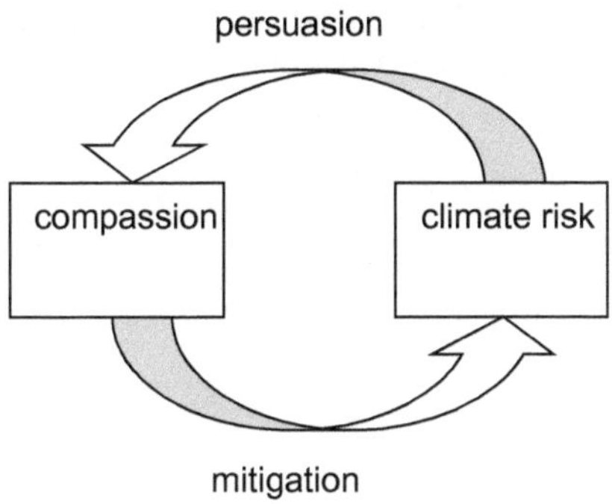

2.1 Compassion

Our pool of compassion rises and falls. It rises when we are paying attention to climate risk; it falls when we are not paying attention to climate risk.

We pay attention to climate risk when we—or more of us—become more persuaded of how it threatens future generations. Persuasion means discovery, explication, inspiration, and argument. Persuasion translates a statistical estimate of the risk of climate catastrophes to images of misery, and then to compassion. Persuasion gathers strength as more scientists step forward to articulate global warming, when activists lie down at gates of coal-burning power plants, when schools hire more teachers of ethics, or when more Al Gore's, more Pope Francis's, and more Obama's speak up for future generations (Broome 2012; Stern 2015).

At the same time, compassion fades. The compassion that affected me so deeply yesterday affects me less today and will affect me even less tomorrow. Like a tire, compassion becomes flat over time. All emotions do.

2.2 Climate Risk

The other pool, "climate risk," also rises and falls. At any moment, its level depends on the balance of two forces: "business as usual" and "action." These two forces reflect our bifurcated mind (Kahneman 2011). On any given day, we go about business as usual, driving to work in SUV's and flying to conferences (IPCC 2014). On the same day, we are mindful of climate risk to future generations, so we email our senators to urge them to regulate carbon emissions from fossil fuels.

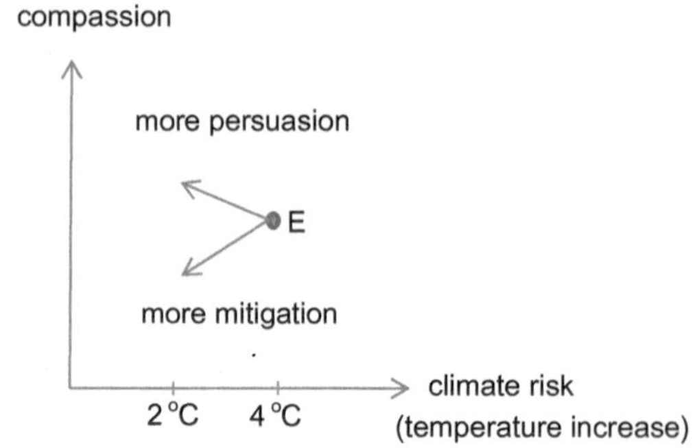

Fig. 2 The emotional bias against persuasion

The upshot is that feedback loop has a steady state, where compassion and climate risk become stable.[1] That steady state hinges on strengths of persuasion and action. Consider Fig. 2. (I measure climate risk by increase in temperature.) The steady state is represented by point E. The arrows emanating from E show what happens when either persuasion or action becomes stronger. When persuasion becomes stronger, E moves northwest; when action becomes more effective, E moves southwest. The difference lies in what happens to compassion: in one case compassion goes up and in the other compassion goes down.

3 The Poverty in Persuasion

In my story so far, both persuasion and mitigation reduce climate risk. Now suppose there is a sum of money to reduce climate risk. How will we proportion the money between persuasion and mitigation?

I believe that we will not invest in persuasion. The reason is that compassion is a negative emotion. Look at Fig. 2 again. There, we have a choice between two investments that reduce climate risk by an equal degree but change our compassion in opposite directions. How would I choose? I would choose the one with *lower* compassion. Since our compassion springs from our perception of others' suffering and misery, compassion is unpleasant. Compassion feels like guilt and shame.

[1]The feedback loop may also hit a tipping point and become chaotic. A "tipping point" is that glaciers in Greenland and Antarctica (Wagner and Weitzman 2015) completely melt away.

We want to avoid that negative feeling. This is one behavioral bias against persuasion.

Another behavioral bias against persuasion is the free-rider behavior. While inventors of green technology reap financial benefits through patents and licenses today, climate activists who specialize in persuasion do not. We treat their services as a free good, hoping that someone else will donate a lot of money for the cause. If persuasion falters and declines as a social force, there is willful blindness to fall back on.

In the least, the art and science of persuasion, though already an established academic field, deserves more attention. Marshall (2014) shows how we can do persuasion by sharing stories and narratives. Stoknes (2015) advocates persuasion that emphasizes the positive rewards for taking actions rather than the negative consequences for not taking actions. Hoffman (2015) urges scientists to become more engaged in persuading the public of their findings. Stern (2015) and Wagner and Weitzman (2015) advocate activism of all kinds, on any scale. Also, see IPCC (2014) for a discussion of how behavioral economics can help promote persuasion.

Here is another thought. Persuasion is like fashion and opinion: it may hit tipping points. When it does, it explodes or implodes. Gladwell (2006) shows us how.

4 A Note on the Treatment of Future People in Economics

As I see it, intergenerational justice is a state of our mind. It is imagined. The reason is simple: as I observed in the beginning of this chapter, future generations are a singular abstraction; they exist only in our mind.[2]

Yet, for the longest time, (most) economists assume that there will always be people in the distant future, that these future people would think and feel like us, and that they will enjoy an increasingly higher standard of living. As a consequence, economists treat something that happens mostly in the future, such as climate change, as an exercise in benefit-cost analysis. Economists love to hunt for the "right" discount

[2] Also see Cowen and Parfit (1992).

rate with which to count the well-being of future generations against our well-being today.

To be fair, I note here that some economists feel uncomfortable about treating Future Earth like investing in a shopping mall. In 1999, twenty of them got together to see if anything could be done about it. The organizers of the conference posed the following question to the participants:

> Perhaps more fundamentally, is it appropriate to use benefit-cost analysis at all in decisionmaking on such issues as climate change, disposal of high-level nuclear wastes, and so on? (Portney and Weyant 1999, p. 4)

Most participants answered, unequivocally, “no.” Solow (1999, p. ix), who later would write the foreword for the proceedings, wondered if there was a “different animal” for how we approach intergenerational justice:

> Maybe the idea of a unitary decision-maker – like an optimizing individual or a wise and impartial adviser – is not very helpful when it comes to the choice of policies that will have distant future effects about which one can now know hardly anything. Serious policy choice may then be a different animal, quite unlike individual saving and investment decisions.

I hope you have noticed that this chapter is a “different animal.”

Appendix: The Compassion Loop in Math

In the text, I refer to “steady state” of the compassion loop. Here I derive it in the simplest mathematical model of the feedback loop. I also show that Fig. 2 is a property of the model.

The Compassion Loop, in Fig. 1, consists of two branches: the evolution of compassion and the evolution of climate risk. These branches are captured by two differential equations:

$$\frac{dS}{dt} = pR - bS \tag{1}$$

$$\frac{dR}{dt} = c - (dR + mS) \tag{2}$$

In the equations,

t	time;
S	compassion;
R	climate risk;
p (>0)	strength of persuasion;
b (>0)	rate of fading of compassion;
c	"business-as-usual" growth in climate risk;
d	Earth's ability to reduce climate risk by breaking down greenhouse gases;
m	strength of mitigation.

We find the steady state by setting the left-hand sides of the equations to zero and then solving them for S and R:

$$S = \frac{c}{\left(\frac{bd}{p}\right) + m} \tag{3}$$

$$R = \frac{c}{\left(\frac{pm}{b}\right) + d} \tag{4}$$

From these expressions, we can tell by inspection how extra strengths in persuasion and mitigation affect the steady state:

(i) A rise in the strength of persuasion (i.e., parameter p) shall raise compassion but lower climate risk, as shown Fig. 2.
(ii) A rise in the rate of mitigation (i.e., parameter m) shall lower both compassion and climate risk, as shown in Fig. 2.

References

Broome, John. *Climate Matters: Ethics in a Warming World*. New York: W. W. Norton, 2012.

Burton, Neel. "Empathy vs Sympathy." Accessed at https://www.psychologytoday.com/blog/hide-and-seek/201505/empathy-vs-sympathy.

Cowen, Tyler, and Derek Parfit. "Against the Social Discount Rate." In Peter Laslett and James S. Fishkin, eds. *Justice Between Age Groups and Generations*. New Haven: Yale University Press, 1992.

Dasgupta, Partha, Karl-Göran Mäler, and Scott Barrett. "Intergenerational Equity, Social Discount Rates and Global Warming." In Paul R. Portney and John P. Weyant, eds. *Discounting and Intergenerational Equity*. Washington, DC: Resources for the Future, 1999.

Dietz, Simon, and Nicholas Stern. "Why Economic Analysis Supports Strong Action on Climate Change: A Response to the *Stern Review*'s Critics." *Review of Environmental Economics and Policy*, 2 (1), 2008, pp. 94–113.

Frank, Robert. *Passions Within Reason*. New York: W. W. Norton, 1988.

Gladwell, Malcolm. *The Tipping Point: How Little Things Can Make a Big Difference*. New York: Little, Brown and Company, 2006.

Hoffman, Andrew. *How Culture Shapes the Climate Change Debate*. Stanford: Stanford University Press, 2015.

IPCC (Intergovernmental Panel on Climate Change). "Social, Economic and Ethical Concepts and Methods." In *Climate Change 2014: Mitigation of Climate Change*. Contribution of Working Group III to the Fifth Assessment Report of the Intergovernmental Panel on Climate Change. Accessed at https://www.ipcc.ch/pdf/assessment-report/ar5/wg3/ipcc_wg3_ar5_chapter3.pdf.

Kahneman, Daniel. *Thinking, Fast and Slow*. New York: Farrar, Straus and Giroux, 2011.

Marshall, George. *Don't Even Think About It: Why Our Brains Are Wired to Ignore Climate Change*. New York: Bloomsbury USA, 2014.

Portney, Paul R., and John P. Weyant, eds. *Discounting and Intergenerational Equity*. Washington, DC: Resources for the Future, 1999.

Sakurai, Misuchi. "Pity, Sympathy, Compassion & Empathy." Accessed at https://empathicperspectives.wordpress.com/2009/05/05/pity-sympathy-compassion-empathy/.

Solow, Robert M. "Foreword." In Paul R. Portney and John P. Weyant, eds. *Discounting and Intergenerational Equity*. Washington, DC: Resources for the Future, 1999.

Stern, Nicholas. *Why Are We Waiting? The Logic, Urgency, and Promise of Tackling Climate Change*. Cambridge: The MIT Press, 2015.

Stoknes, Per Espen. *What We Think About When We Try Not to Think About Global Warming: Toward a New Psychology of Climate Action*. White River Junction: Chelsea Green Publishing, 2015.

Wagner, Gernot, and Martin Weitzman. *Climate Shock*. Princeton: Princeton University Press, 2015.

Index

L. W. Lee, *Behavioral Economics and Bioethics*,
Palgrave Advances in Behavioral Economics,
https://doi.org/10.1007/978-3-319-89779-0